PETROCHEMICALS
IN NONTECHNICAL LANGUAGE

PETROCHEMICALS

IN NONTECHNICAL LANGUAGE

DONALD L. BURDICK

WILLIAM L. LEFFLER

PENNWELL PUBLISHING COMPANY
TULSA, OKLAHOMA

PennWell Publishing Company
1421 South Sheridan/P.O. Box 1260
Tulsa, Oklahoma 74101

Library of Congress Cataloging-in-Publication Data

Leffler, William L.
Petrochemicals in non-technical language / W.L. Leffler, D.L. Burdick.
p. cm.
ISBN 0-87814-344-0
1. Petroleum chemicals. I. Burdick, Donald L. II. Title.
TP692.3.L44 1990
661'.804--dc20 90-7062
CIP

Printed in the United States of America

2 3 4 5 94 93 92 91

Cover photo; Quantum Chemical Corp. plant, Tuscola, Illinois.

CONTENTS

PREFACE

This new, improved, renamed edition is designed to be used in any of five ways:

- Read it cover to cover for a nontechnical education covering 90% (by volume) of the traded petrochemicals. There are even exercises at the end of each chapter to test comprehension and retention. Complete answers are in the back of the book.
- Read a chapter or section as the subjects come up in your business life. Each one is designed to be a self-contained description of one petrochemical. If you're too busy, there's a 10 sentence summary at the end of each chapter.
- Use it as a nontechnical encyclopedia. The index in the back is extensive, and the primary coverage of each entry is highlighted in bold page numbers.
- Use it as a primer in petrochemical economics. There are material balances in many chapters. A number of the exercises deal with product or process economics.
- Recommend it to your subordinates, colleagues, or your superiors who need to know at least half as much about petrochemicals as you do.

There are five sections in this book. The first is only one chapter—it's the mandatory discussion of chemistry. Our editors tell us the book would not be technically complete without it. It's not bad, but some readers have been known to skim/skip it.

The next section covers the building blocks, from which most of the remaining petrochemicals are derived. Then there are two sections on the first line derivatives. The grouping within these two sections is for convenience mostly.

The last section is on polymers, which are "borderline" petrochemicals. We debated whether they belong in a book about petrochemicals, but then we wrote them and they seemed to complete the linkage from raw materials (coal, oil, gas) all the way to consumer products. If you don't agree, don't read them.

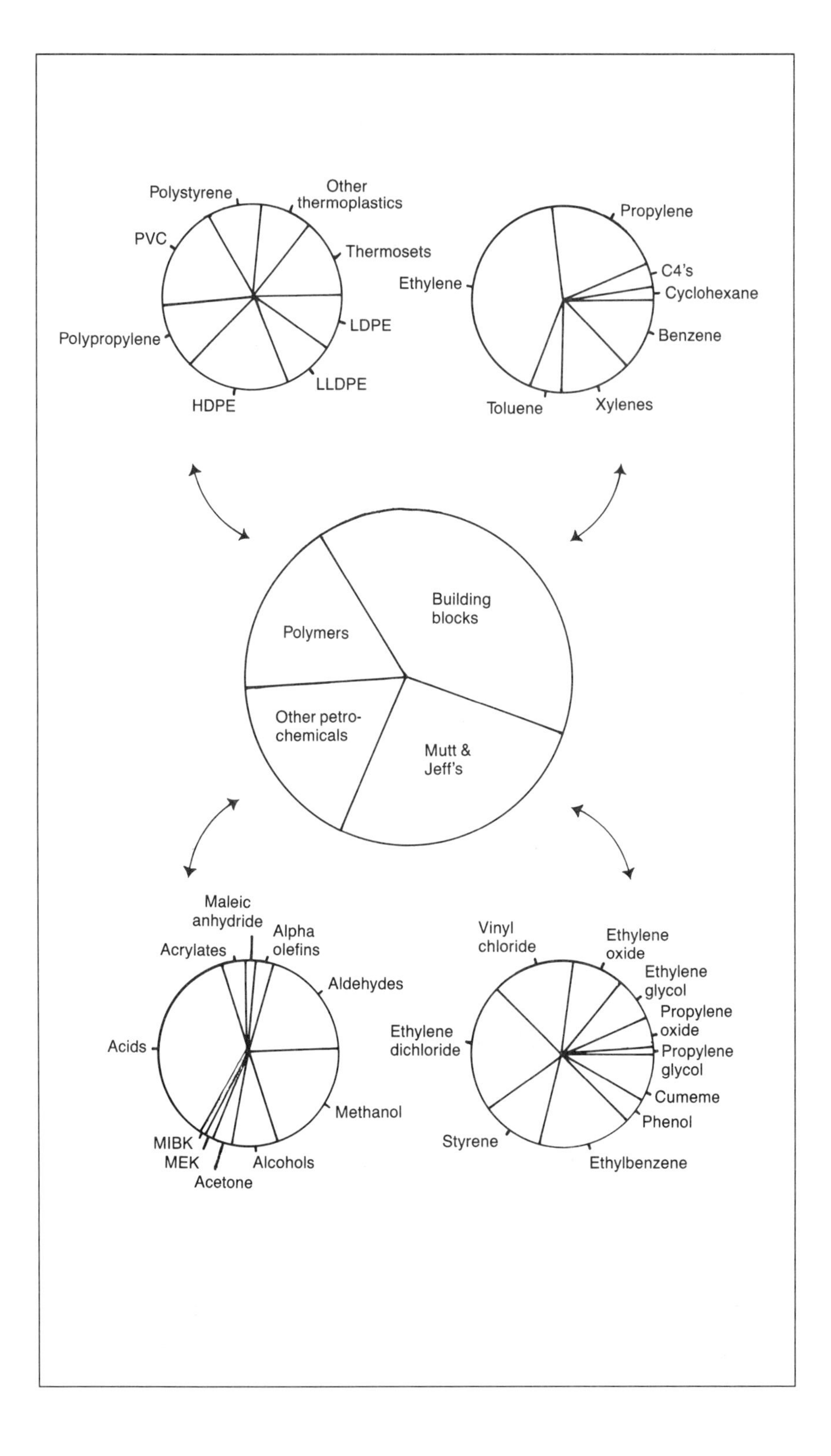
Polystyrene
Other thermoplastics
PVC
Thermosets
LDPE
Polypropylene
LLDPE
HDPE
Propylene
C4's
Ethylene
Cyclohexane
Benzene
Toluene
Xylenes
Building blocks
Polymers
Other petro-chemicals
Mutt & Jeff's
Maleic anhydride
Alpha olefins
Acrylates
Aldehydes
Acids
Methanol
MIBK
MEK
Acetone
Alcohols
Vinyl chloride
Ethylene oxide
Ethylene glycol
Propylene oxide
Ethylene dichloride
Propylene glycol
Cumeme
Phenol
Styrene
Ethylbenzene

For perspective, the relative volumes of the petrochemicals that are covered in this book are shown in Figure P–1. There is some redundancy between the building blocks, the derivatives, and the polymers because the latter are made from the former, but the redundancy is meant to be included in the diagram for perspective.

Through the wonders of text scanners, personal computers, and word processing, the authors have taken care of this new edition themselves. We even read the entire manuscript aloud to each other, in the fashion James Michener does with his epic novels. So we have no one to blame for errors or omissions but each other. So be it.

D. L. B.
W. L. L.

I

WHAT YOU NEED TO KNOW ABOUT ORGANIC CHEMISTRY

"The time has come," the Walrus said, "to speak of many things:
Of shoes—and ships—and sealing wax—of cabbages—and kings."

■

Through the Looking Glass
Lewis Carroll, 1832–1898

What is organic chemistry? It's the study of compounds containing carbon, and it's fundamental to understanding petrochemicals. Why the word organic? you might ask. Originally, and that means before 1800, organic was applied only to compounds whose formation was supposed to be due to some living force such as plants or animals. Then early in the nineteenth century, a chemist named Wohler synthesized urea, the main ingredient in urine. (History didn't record why he was trying to do that.) Up until that time it was believed urea could only be produced "organically," by animal life. Therefore, and until today, the term organic chemistry was stretched beyond its original meaning to include all carbon compounds. So now the difference between organic and inorganic chemistry is more definitional than natural.

It may be surprising to find out that organic compounds comprise more than 95% of all compounds known to exist, and that's over a million. Three things about carbon and carbon compounds help explain the proliferation of organic chemicals. The first is the electronic configuration of the carbon element. Don't leave now . . . you're about to get the 2.2 minute summary of the periodic table of elements, atoms, electrons, protons, valences, bonds, and compounds.

There are about 100 different kinds of atoms that make up all kinds of matter, and they can be classified in a table—the Periodic Table of Elements—according to their construction. In the center of any atom is a nucleus containing protons and neutrons. The protons have a positive charge, the neutrons are neutral. So the nucleus is positively charged. Electrons, equal in number but opposite in charge to the protons, move around the nucleus in orbits. You might think of an atom like a solar system. The nucleus acts like the sun; the electrons orbit the nucleus like the planets circle the sun.

There is one difference, however. The innermost orbit can contain either one or two electrons, at most. The next orbit can have up to eight electrons. The succeeding orbits become a more complex story, but luckily almost all petrochemicals have atoms with no more than two orbits.

The rules of electrons and orbits are important because the number of electrons in the outermost ring determines some of the more important chemical properties of the element. Atoms have a yearning to move towards maximum stability by filling up their outermost orbit to the maximum content. Atoms can gain or shed electrons, or share them with another atom in the process of achieving the stability of a complete set of two or eight electrons. For example, take the carbon atom. It has six neutrons and six protons in the nucleus and six electrons in orbit. The first orbit has two; the second has the required remaining four. These four are called the valence electrons. Carbon has a valence of four because it needs four more electrons to fill the outer ring up to its capacity of eight. It desperately wants to find some other atoms with which it can share four electrons.

Another example is hydrogen. Hydrogen has one proton and one neutron in the nucleus and one electron in the first and only orbit. It needs another electron in that orbit to stabilize itself. Figure 1-1 shows how carbon and hydrogen can achieve mutual satisfaction in the marriage of two atoms into a compound, methane.

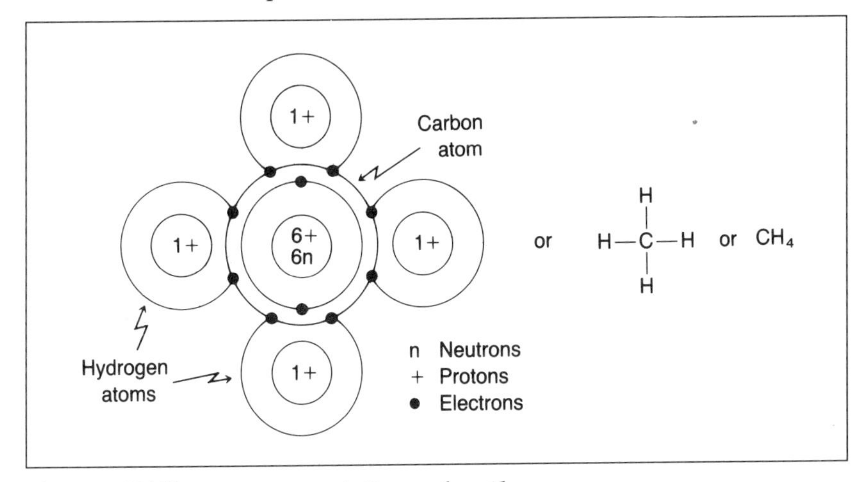

Fig. 1–1 Different representations of methane

Each of the four hydrogen atoms share their one electron with the carbon atom to satisfy carbon's need for eight electrons in its outer ring. The carbon shares an electron with each hydrogen atom to satisfy the need for two electrons in the hydrogen atoms' outer ring. The polygamous result is a stable compound with all the proton and electron charges balanced.

Carbon and hydrogen can even link up with other carbons and hydrogens. When hydrogen hooks up with another hydrogen, it forms H_2, and the electron "needs" of the two hydrogen atoms are satisfied. But when carbon hooks up with another carbon, each carbon still has a need for three more electrons. Filling them out with hydrogens is possible, and when that happens, the compound ethane forms, as shown in Figure 1-2.

```
    H   H
    |   |
H—C—C—H
    |   |
    H   H
```

C_2H_6

Fig. 1–2 Ethane

This is a long way to explain that the propensity of carbon to connect with four other atoms is the partial explanation of why there are so many carbon compounds: there are lots of ways atoms can hook up with carbon atoms.

The second characteristic unique to carbon compounds that also helps explain why there are so many is called isomerism. Compounds with the same number and kinds of atoms can have very different properties. For example, glucose has the formula $C_6H_{12}O_6$. Yet there are 16 other compounds with the same number of carbons, hydrogens, and oxygens. Its not likely, though, that you'd like your night nurse to hook up galactose or fructose to your intravenous instead of glucose, even though they have the same formula. The difference is that the atoms are linked together in such a way as to have different spatial configurations, and, as you'll see, that makes them behave differently, physically and chemically. Such similar but different compounds are called isomers.

So if you put the phenomenon of isomerism together with the propensity of carbon to react (the valence of four), and add to that Mother Nature's blessing of an abundance of carbon in this universe, you can understand the preponderance of compounds and the importance of organic chemistry.

One further characteristic unique to carbon is important and needs to be covered before leaving the subject of valences: bonds. A few paragraphs ago you learned that carbon could link up to itself and three other atoms. In fact, carbon can also link up to itself with double bonds or triple bonds to "satisfy" its valence requirements of four. For example, in Figure 1–3 two carbon atoms are linked together with single, double, or triple bonds filled out with hydrogens, forming three different compounds: ethane, ethylene, and acetylene.

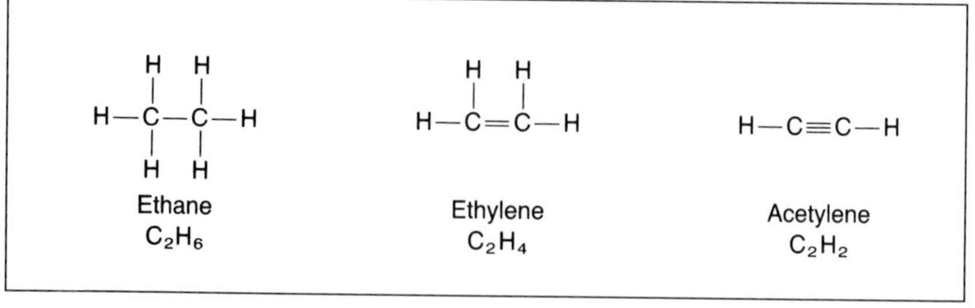

Fig. 1–3

You will find that in petrochemical processes, the more multiple bonds, the more reactive the compound is. Acetylene is much more likely to react with other compounds (explosively, in fact) than ethylene, which itself is far more reactive than ethane. You can think of the double or triple bonds as squeezing into the place suitable for one bond. As a result there is pent up pressure to relieve the stress and increased chemical reactivity.

As a matter of common nomenclature in the petrochemical world, at least when you hear chemical engineers or chemists talking, carbon compounds with single bonds are called saturates. (The carbon atoms are saturated with other atoms.) Those with multiple bonds are called unsaturates. Double bond reactivity is characteristic of the basic building blocks of the petrochemical business. Ethylene, for example, is the chemical compound used to make vinyl chloride, ethylene oxide, acetaldehyde, ethyl alcohol, styrene, alpha olefins, and polyethylene, to name only a few.

Going on from here can be a morass if there's no organization, so you need to look at one of the generally accepted breakdowns of organic chemicals, shown in Figure 1–4.

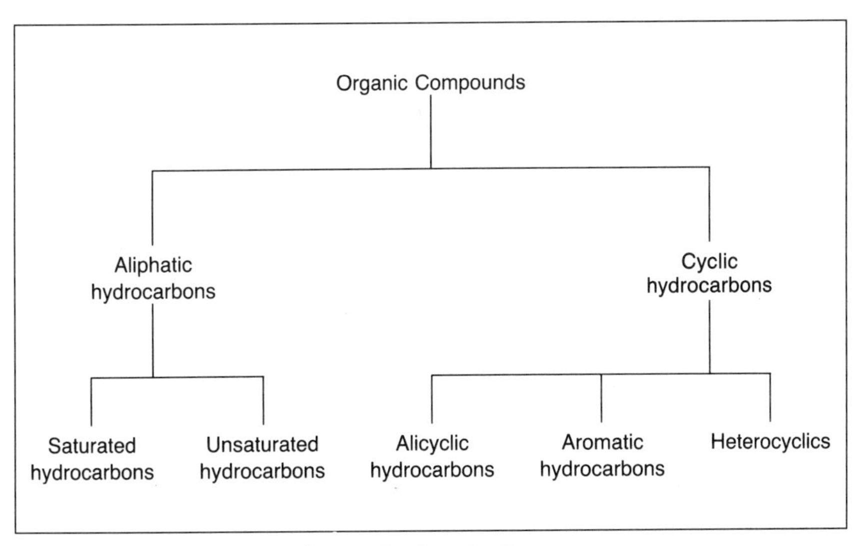

Fig. 1–4 Classification of organic chemicals

The aliphatic hydrocarbons have already been introduced . . . they contain only hydrogen and carbon atoms; they can have single or multiple bonds—they're saturated or unsaturated hydrocarbons. The simplest member of the group is methane, and more complicated molecules (combinations of atoms) in this group can be formed by adding additional combinations of one carbon with two hydrogens attached to it between any carbon and hydrogen atoms like in Figure 1–5.

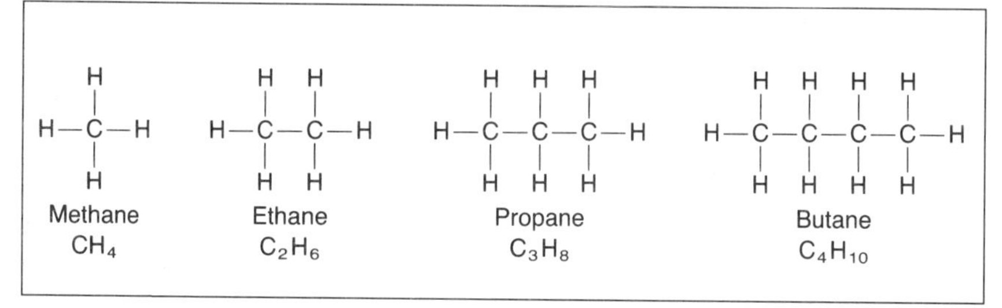

Fig. 1–5 Paraffins

The whole family that results from endless addition of the -CH_2's is called the paraffin series. The word comes from the name of the wax little old ladies used in "the old days" to seal jelly jars. That particular paraffin consists of a mixture of $C_{30}H_{62}$'s on up to $C_{50}H_{102}$'s. Note that the formulas always have twice as many hydrogens plus two, compared to the carbons. That's the way it works out.

The petroleum products processed in oil refineries are predominantly paraffins, and they are often characterized by the temperature at which they boil. Distillation or fractionation, one of the most useful processes in refining, is based on these boiling points. For example, at room temperatures, the following petroleum-type paraffins take the three basic forms of matter we see in nature—gas, liquid, and solid:

CH_4, C_2H_6, C_3H_8, and C_4H_{10} are gases but they liquify at or below 32°F (0°C);

C_6H_{14} through C_9H_{20} are liquids, but they boil between 150 and 300°F (65°C and 150°C);

$C_{30}H_{62}$ and bigger are solids, but they melt at 200–300°F and boil above 500°F (260°C).

Unsaturated hydrocarbons are typified by ethylene. Propylene, butylene, and bigger molecules are structured in the same manner as the saturates, but one of the single bonds is replaced with a double bond, as shown in Figure 1-6. Another popular name for these compounds is olefins.

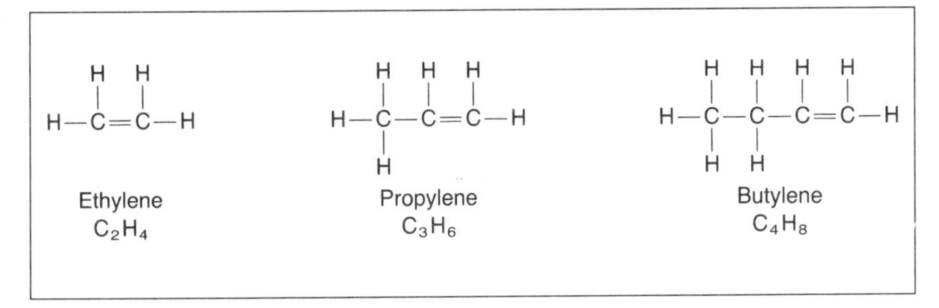

Fig. 1–6 Olefins

The double bond difference between the olefins and the paraffins is the essence of the difference between the petrochemicals and petroleum products—the former depend much more on the chemical reactivity of the double bonded molecules. While paraffins can be manipulated in refineries by separation or reshaping, olefins in a petrochemical plant are usually "reacted" with another kind of atom or compound such as oxygen, chlorine, water, ammonia, or even with more of itself. The result is more complicated compounds useful in an increasing number of chemical applications. More on this in later chapters.

Stop for a little interlude to pick up two auxiliary, but important, concepts. The first is organic groups. The other is isomers, which should have a familiar ring.

Organic group is a handy chemical shorthand notation for a cluster of atoms that looks much like the stand-alone molecule after which it is named. Take the methyl group . . . it's nothing more than methane with one of the hydrogens missing, as shown in Figure 1-7. But it is attached to some other atoms to make up a larger molecule, methyl alcohol. Organic groups are not stand-alone molecules themselves. They are always part of a molecule.*

*Often the terms organic groups and radical are incorrectly used interchangeably. A radical looks like an organic group, except it can stand alone, unattached to a molecule. As a result it has an unpaired, odd electron and is extremely reactive. The methyl radical $\bullet CH_3$ can be produced, with some effort, from methane by the loss of one hydrogen atom. The use of the letter R, which is the first letter of the word radical, for organic compounds doesn't help dispel the confusion between the two terms. (Even this explanation had to be written twice before it sounded right.)

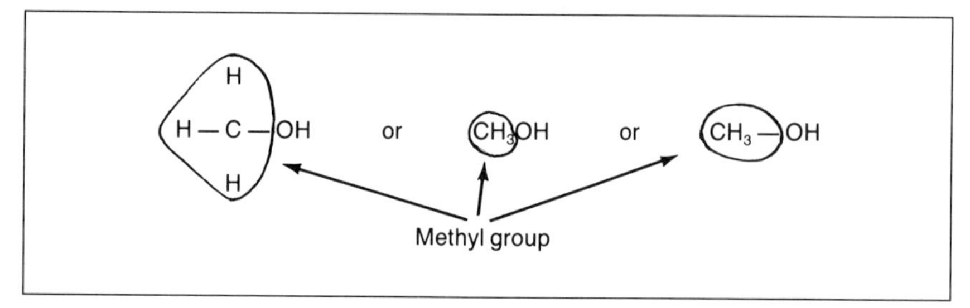

Fig. 1–7

The shorthand symbol for organic groups is R-. Technical writers (and chemistry teachers) use R- whenever they want to indicate that any number of organic groups could be attached to make a molecule. In Figure 1-7, the methyl group, $-CH_3$, could be represented by R-.

Another good example of organic groups is shown in Figure 1–8, a diagram of tetraethyl lead. This is the additive that was put in gasoline to improve the octane rating. Tetraethyl lead has four ethyl groups ($-C_2H_5$) attached to the element lead (Pb).

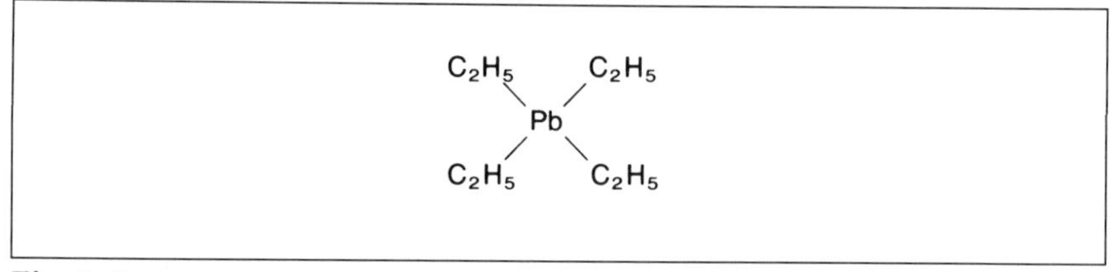

Fig. 1–8

For the most part, in the rest of this book, the organic group notations in the middle and at the right side of Figure 1–7, will be used. They are a lot less clumsy than the sprawl at the left. Occasionally, the notation R- will turn up too.

Now back to isomers. To firmly implant them in your mind, consider butane and its isomer, iso-butane, in Figure 1-9. The difference between the two C_4H_{10} molecules is how the organic groups $-CH_3$ are connected. In iso-butane, one of the carbons has three, not two, methyl groups attached to it. But there is more to the difference between the two molecules than just drawing them. Iso-butane behaves differently as well. It boils at a different temperature, it gives off a different amount of heat when it burns, has different chemical reactivity, and so on.

The butylene isomers shown in Figure 1–10 add another degree of complexity because of the double bond. It is an easy mistake to go overboard in drawing isomers that have the same formula but appear to look different. But be careful, because molecules don't know left from right or front from back. What may look different on paper may be identical when rolled over in space. That's why the iso-butylene in Figure 1–10 is drawn the way it is. If you try to attach that =CH2 group to some other carbon in the molecule, the whole thing becomes a normal butylene.

Like the butane isomers, the butylenes each have their own properties which make them unique, and of individual appeal to the petrochemical industry.

The root *-mer* figures importantly in petrochemical nomenclature. It comes from the Greek word meros, which means part. The chemists picked up its usage to define how organic groups are linked together. You will find it imbedded in the following:

monomer (with mono-, one)—a compound capable of reacting with itself or other similar compounds.

dimer (with di-, two)—two monomers joined together, e.g., butene.

trimer (with tri-, three)—three monomers joined together, e.g., hexene.

oligomer (with olig-, a few)—up to ten, more or less, monomers joined together in a string, e.g., alpha olefins.

polymer (with poly-, many)—multiple monomers linked together, e.g., polyethylene.

isomer (with iso-, equal)—molecules with an equal number and kind of atoms arranged differently e.g., butylene and iso-butylene.

Dimers, trimers, and tetramers are all forms of oligomers.

CYCLIC COMPOUNDS

The fundamental difference between cyclic compounds and the others already covered is the arrangement of the carbon atoms in a cyclic structure. Cyclic compounds have a closed chain of carbon atoms. Cyclopropane, shown in Figure 1–11, is the simplest cyclic hydrocarbon.

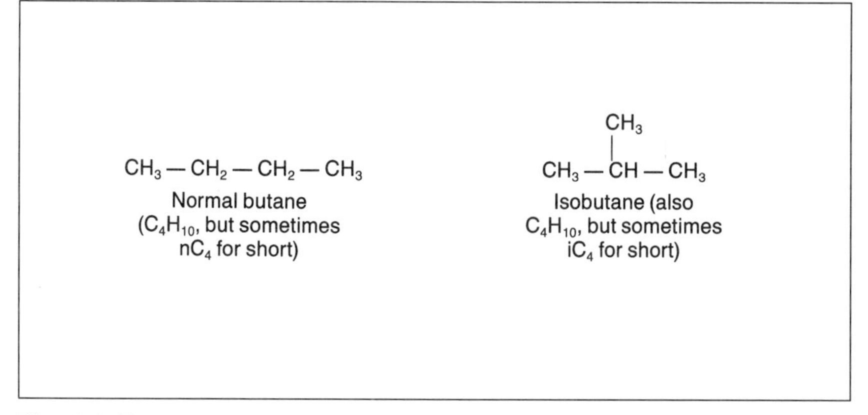

Fig. 1–9 Butanes

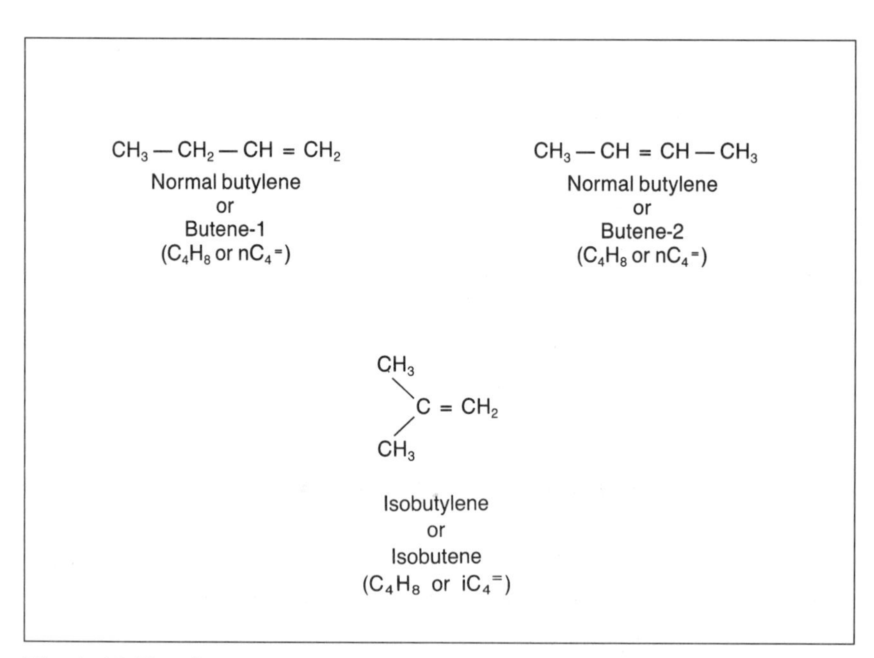

Fig. 1–10 Butylenes

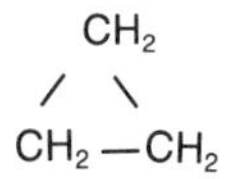

CH_2
/ \
CH_2 — CH_2

Fig. 1–11 Cyclopropane

You may have been administered cyclopropane in a dentist's chair—it's one of the several anesthetics used to put you to sleep. Others are nitrous oxide (laughing gas) and ether which you will run into later on.

Cyclopropane, and the cyclic compounds shown in Figure 1–12, cyclopentane and cyclohexane, are members of the alicyclic branch shown in Figure 1–4. The ali- is the same prefix as used in the aliphatics because of the structure. Except for the cyclic formation, they are made up basically of chains of methylene groups ($-CH_2-$). But one difference from the aliphatic series of organics is the chemical reactivity. Lower members of the alicyclic series have one chemical property similar to double-bonded olefins—they are quick to react chemically. The simple explanation for this reactivity is that the bonds attaching the carbons to each other are strained because of the angles they must take. They're bent out of their "natural" shape. In any chemical reaction, the rings readily open up to alleviate this strain.

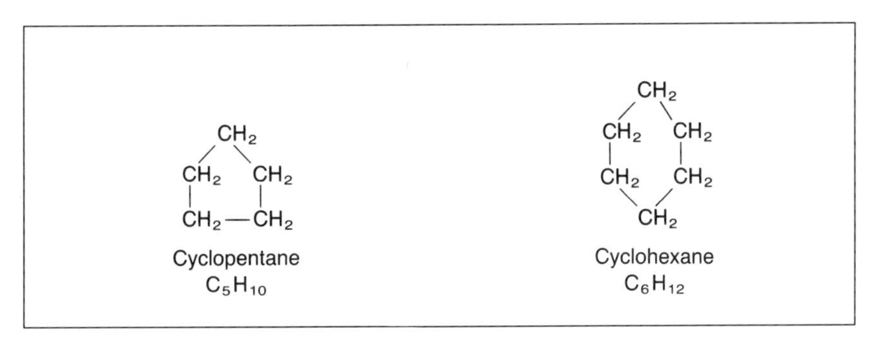

Fig. 1–12 Alicyclics

You might surmise that if there are more carbon atoms in the ring, the compounds might be more stable. In fact cyclopentane and cyclohexane are much more stable, and like the paraffins, slower to react. They'll burn, of course, but not explosively.

Cyclopentane and cyclohexane are commonly found in petroleum products like gasoline and are generically called yet another name, naphthenes, in the refining business.

AROMATICS COMPOUNDS

By far the most commercially important family of compounds on the cyclic side of the road map in Figure 1–4 are the aromatic compounds. Benzene is the patriarch. Like much of the nomenclature in organic chemistry, the term *aromatic* is a misnomer. It is a legacy from the nineteenth century when a group of unsaturated compounds of high reactivity and with a sickly, sweet, hydrocarbonish smell were isolated and fell under the name aromatics. Unlike the term organic, which got broader meaning in the twentieth century, the name aromatics got narrower and is limited today to benzene and benzene derivatives.

The benzene molecule is a remarkable structure with six carbons in a hexagonal ring. To satisfy the carbon valence of four, every other carbon-to-carbon link is a double bond, and each carbon has only one hydrogen attached.

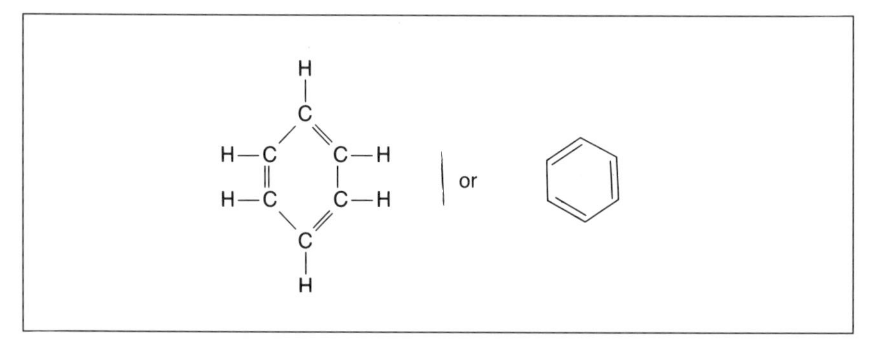

Fig. 1–13 Benzene, C_6H_6

There are some subtle but very important characteristics unique to the benzene ring. One is symmetry. Every carbon in the ring looks like every other carbon; every hydrogen looks like every other hydrogen. There are no benzene isomers. Every benzene molecule looks like every other benzene molecule. Moreover, as in many chemical reactions covered in later chapters, when one of the hydrogens is replaced during a chemical reaction, resulting in something called a monosubstituted benzene, that compound also has no isomers. The beneficial fallout of this phenomenon is that the products of monosubstitution are identical, homogeneous. Take the compound toluene, for example. In Figure 1–14, all three molecules are benzene with a methyl group replacing a hydrogen. While they appear to be different, each needs to be rotated just a little to make it look like the others. So there's really only one kind of toluene, as toluene is a monosubstituted benzene.

The next logical step is to replace two hydrogens on the benzene ring (di-substitution). Three isomers occur. Take xylene, for example, $C_6H_4(CH_3)_2$. That's a benzene ring with methyl groups replacing two hydrogens. As you can see in Figure 1–15, the replacement can be in one of three patterns (and only three, if you look closely.)

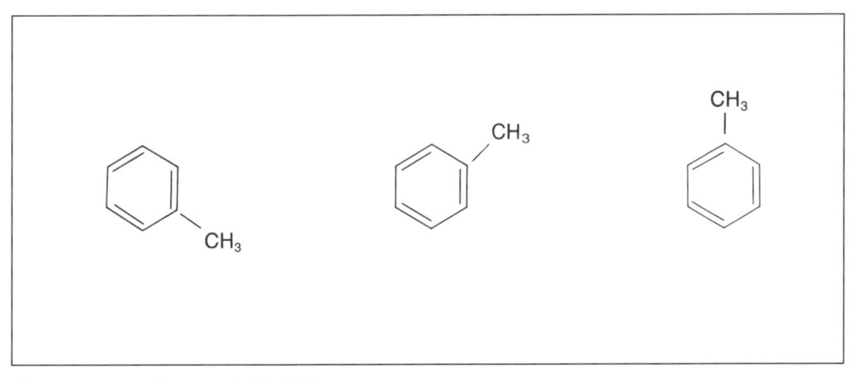

Fig. 1–14 Toluene $C_6H_5CH_3$

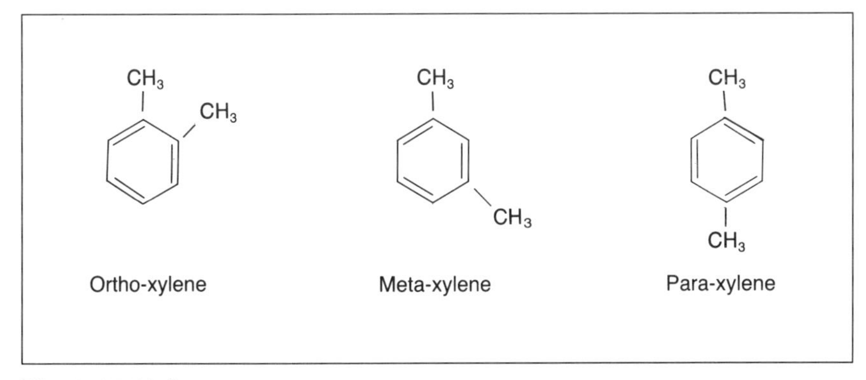

Fig. 1–15 Xylenes

The isomers, called ortho-xylene, meta-xylene, and para-xylene, each have unique properties. Two such properties are the freeze points, at which xylenes turn from liquid to crystals, and the boiling points, at which xylenes turn from liquid to vapor. These two properties figure importantly in the apparatus used to separate xylene isomers from each other. Mixed xylenes, a commonly traded commodity, is a combination of the three isomers.

Di-substituted benzenes like xylenes might be compared to a litter of puppies. They're all dogs, but each one behaves differently. If they were mono-substituted benzenes like toluene, they'd all be clones.

HETEROCYCLICS

Take one of the alicyclics or aromatic compounds which have a chain of carbon atoms in a closed ring, and replace one of the carbon atoms with some other kind of atom (usually it's oxygen, nitrogen, or sulfur), and you have a heterocyclic compound. Ethylene oxide is the simplest of the heterocyclic series since it is a three atom ring. (Anything smaller wouldn't be a ring.) Ethylene oxide is shown in Figure 1-15. The oxygen atom doesn't have any hydrogen atoms attached because it has a valence of two. The bonds with the carbon atoms satisfy the oxygen valence requirement.

```
      O                      O
     / \                    / \
  CH2 — CH2              CH2 — CH — CH3
```

Ethylene oxide, $CH_2\,OCH_2$ Propylene Oxide, $CH_2\,OCH\,CH_3$

Fig. 1–16 Cyclic Oxides

Propylene oxide, another commercially important chemical, also is shown in Figure 1–16, illustrating two concepts already discussed—heterocyclics and a methyl group replacing a hydrogen atom.

EVERYTHING ELSE

The list of "everything else" is expanding endlessly and could be the longest section of this chapter. University students take numerous advanced courses to learn about them. Mercifully, this section of "everything else" is limited to brief discussion of the few classes of compounds that have become important in the petrochemical business.

You are now off the Figure 1–4 road map, making organic compounds by adding atoms other than carbon and hydrogen to the Figure 1–4 compounds. The idea is the same as heterocyclics. And as in Figure 1–16, one of the most important of these elements is oxygen, as shown in Figure 1–16.

Oxygenated Organic Compounds

In Table 1–1 there are seven types of organic compounds that have gone through an oxidation process of some kind. That is they have had oxygen chemically added to them in some fashion. Each of these groups of compounds will be discussed in detail in one or more chapters later on, but some familiarity with the nomenclature at this point is helpful. (In each of the chemical formulas in the table, the letter R is meant to represent some organic group or compound in the manner explained above. More importantly, attached to the R is the signature of the family group.)

The characteristic of the alcohols is the addition of the hydroxyl group -OH, the alcohols' personal and unique signature, to another group like the methyl group to make methyl alcohol.

The ketones have an imbedded signature, a carbon atom with a double bonded oxygen attached. Acetone (finger nail polish remover) is the most common ketone.

Aldehydes have a tail end consisting of a carbon/double-bonded oxygen and a hydrogen, both attached to the same carbon. A big volume commodity aldehyde is acetaldehyde. A better known one is formaldehyde.

The acid signature is just a bit more complicated. It's a double-bonded oxygen plus an hydroxyl (-OH) group, both attached to the same carbon. The main ingredient in vinegar, acetic acid, is an example.

Ethers are simple. They have an imbedded oxygen connecting two organic groups which may or may not be identical. Diethyl ether is

Table 1–1 Oxygenated Hydrocarbons

Family Classification	Generic Formula	Examples
Alcohols	$R-OH$	methyl alcohol CH_3-OH; ethyl alcohol C_2H_5-OH
Ketones	$R-C(=O)-R$	acetone $CH_3-C(=O)-CH_3$
Aldehydes	$R-C(H)=O$	acetaldehyde $CH_3-C(H)=O$
Acids	$R-C(=O)-OH$	acetic acid $CH_3-C(=O)-OH$
Esters	$R-C(=O)-O-R'$	ethyl acetate $CH_3-C(=O)-OC_2H_5$
Ethers	$R-O-R$	dimethyl ether CH_3-O-CH_3
Anhydrides	$R-C(=O)-O-C(=O)-R$	acetic anhydride $CH_3-C(=O)-O-C(=O)-CH_3$

the one they give you just before they take your appendix out.

Esters and Anhydrides defy simple explanations, so just look at Table 1-1. The most common ester you have probably encountered is methyl acetate, the solvent put in cans of fast drying, car spray paint. There aren't any commonly used anhydrides around the house.

Nitrogen-Based Organic Hydrocarbons

Just outside the family of organic compounds (like next door neighbors) is a family of compounds based on nitrogen. There are three main branches of the family, shown in Table 1-2. The amines are predominant and generally are formed by reactions involving ammonia, NH_3.

Usually the amines have in them organic groups commonly found in the petrochemical industry. (That's what happens in neighborhoods.) Aniline ($C_6H_5NH_2$), a typical example, is an important dye intermediate.

The nitro compounds are organic compounds linked with the grouping $-NO_2$. Usually the $-NO_2$ comes from nitric acid, HNO_3, like in the reaction of nitric acid and toluene to make trinitrotoluene, TNT.

Finally, a fairly narrow family of near-petrochemicals are the nitriles, compounds with the signature -CN. The family success in this house is acrylonitrile, a compound used extensively in the manufacture of tires, plastics and the kind of fibers that go into sweaters (Orlon and Acrylon).

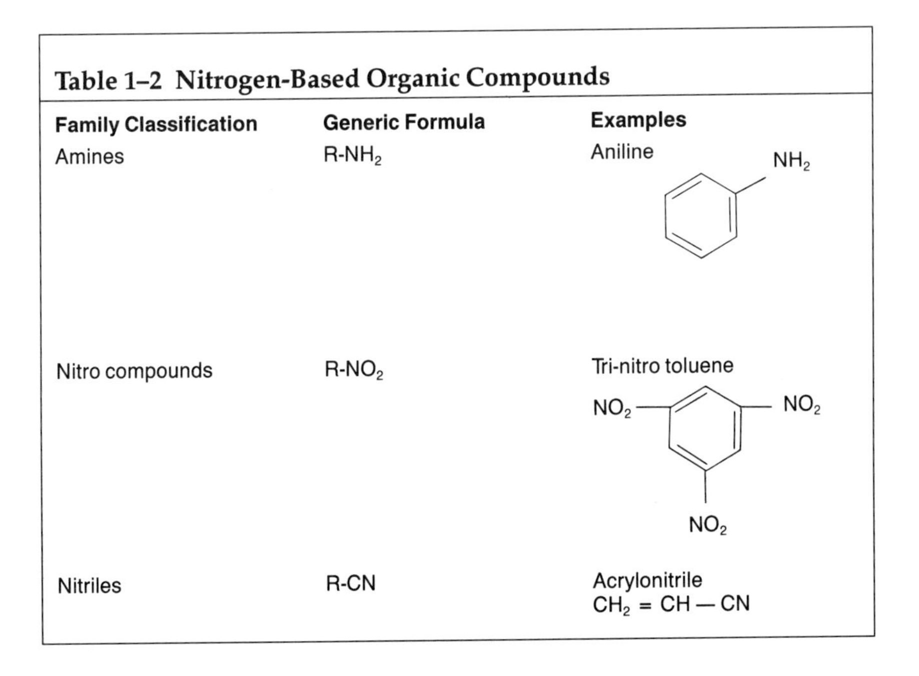

Table 1–2 Nitrogen-Based Organic Compounds

Family Classification	Generic Formula	Examples
Amines	$R-NH_2$	Aniline (benzene ring with NH_2)
Nitro compounds	$R-NO_2$	Tri-nitro toluene (benzene ring with three NO_2 groups)
Nitriles	R-CN	Acrylonitrile $CH_2 = CH - CN$

That's it! This is, by far, the toughest chapter in the book. If you've gotten this far, the rest will be easy. There will be more chemistry as you go along in each chapter, but the doses are small, easy to swallow. So move on!

Exercises

1. Match the items in the left column with the correct corresponding items in the right.

paraffins	benzene, xylene, and toluene
olefins	paraffins
aromatics	ortho-, para-, and meta-xylene
saturates	C_nH_{2n+2}
unsaturates	C_nH_{2n-6}
isomers	C_nH_{2n}
cyclics	butylenes

2. How many isomers of pentane, C_5H_{12}, are there? Draw them.

3. An Ethyl Group is:
   a. a selection of three types of leaded gasoline
   b. $-CH_3$
   c. $-C_2H_5$
   d. $-C_2H_6$
   e. four little old ladies with the same name

4. If you've got a headache by now, you might take some acetyl salacylic acid. That's aspirin, and it has the chemical structure:

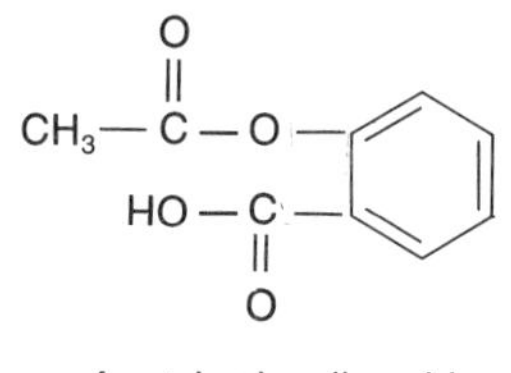

Acetyl salacylic acid
(aspirin)

Find some examples of a methyl group, an ester group, a benzene ring, and an acid group in this molecule.

BENZENE

"Oh my how many torments
lie in the circle
of a . . . ring"

■

The Double Gallant,
Colley Cibber 1671-1757

Why start out with benzene? The obvious answer is that benzene is one of the handful of basic building blocks in the petrochemicals industry, along with ethylene, propylene, and a few others. The more subtle reason is that benzene, more than any of those other chemicals, comes from a broader base—steel mill coking, petroleum refining, and olefins plants. For that reason, the benzene "network," the sources and the uses, is more complex than any of the others.

After a little historical backgound, this chapter will cover benzene production (including the hardware), some of the important properties from the chemist's point of view, and the major benzene applications.

HISTORICAL PERSPECTIVE

Benzene was first isolated and identified in 1825 by Michael Faraday during his scientific heydays at the Royal Institute in London. But benzene proved to be an enigma to chemists for more than a century after that. The valence rules of carbon and hydrogen require that benzene molecules have those characteristic alternating single and double bonds in the carbon ring. But the benzene molecule doesn't behave in the precise way that other molecules with double bonds do. In chemical reactions, the carbon-to-carbon bonds in the benzene ring act in some ways like an average of single- and double-bonded carbons. So in 1865 the German scientist August Kekule offered a very appealing theory. He suggested the single and double bonds continuously trade places with each other—they oscillate or resonate. In the early 1930s, the famous Linus Pauling offered more convincing evidence supporting Kekule's theory, using quantum mechanics. There are still some loose ends, but no good alternate theory has turned up yet.

Benzene had a limited commercial value during the nineteenth century. It was used primarily as a solvent. In the twentieth century, gasoline blenders discovered that benzene had good octane characteristics. As a consequence, there was a large economic incentive to recover all the benzene that was produced as a by-product of the coke ovens in the steel industry. (See below.) Starting around World War II, chemical uses for benzene emerged, primarily in the manufacture of explosives. Not only did the coke-oven benzene get diverted from gasoline blending to the chemical industry, but by mid-century, the refining industry itself was recovering huge quantities of the benzene from gasoline blending stocks to keep up with chemical needs. There's some irony in the fact that the largest consumer of benzene, the petroleum industry, ultimately turned out to be the largest supplier.

The increasing demands for benzene by the petrochemicals industry led to new and improved manufacturing processes—catalytic reforming and toluene dealkylation, a technique for converting toluene to benzene. That process goes in and out of vogue as the economic winds blow to and fro. A fortuitous contribution started in the 1970s when olefin plants started using heavy gas oil as a feedstock and produced by-product benzene.

BENZENE FROM COAL

An important raw material used in the manufacture of steel is coke, a nearly pure form of carbon. To supply themselves with coke, steelmakers developed the process of destructive distillation of coal.

The chemical makeup of coal is predominantly a mixture of very high molecular weight polynuclear aromatic compounds. That mouthful is a common expression used in describing heavy hydrocarbon compounds. High molecular weight refers to the number of atoms, in this case carbon and hydrogen, attached together in the molecule. Ethane, C_2H_6, would be low molecular weight; $C_{30}H_{30}$ would be high molecular weight. Polynuclear aromatic refers to the preponderance of C_6 type rings in the molecule, as you can see in Figure 2–1.

Because of the size of these molecules and the multiple ring feature, the ratio of carbon to hydrogen is high, compared to other hydrocarbons encountered up to this point. In ethane it's 1:3; the compound in Figure 2–1 is almost 2:1. In the destructive distillation process, the coal is heated to 2300–2700°F in the absence of air. At those temperatures, the large molecules begin to crack, forming on the one hand, smaller organic compounds—many of which are liquids or gases at room temperature—and on the other hand pure carbon . . . coke for the steel furnaces.

Because of the high carbon/hydrogen ratio, one ton of coal yields about 1500 pounds of coke and about 500 pounds of coal gas, coal oil, and coal tar. Prior to the advent of electricity, coal gas was the primary source of municipal lighting, and gas lights lined the streets of the great cities in 1900. Coal tar is a solid at room temperature and is often used as a roofing material or for road paving.

The coal oil is a mixture of benzene (63%), toluene (14%), and xylenes (7%), resulting directly from the benzene ring remaining intact during the cracking process. For this reason, steel companies became important suppliers of BTX's (benzene, toluene, and xylenes). Not that they were interested in getting into the chemical business, just that they exploited a valuable by-product.

Benzene from coal coking started to become less important in the 1950s as the market mushroomed and the marginal supply came from petroleum refining. Coal based benzene dropped from nearly 100% in 1955 to 50% in the 1960s and less than 5% in the 1970s and 1980s.

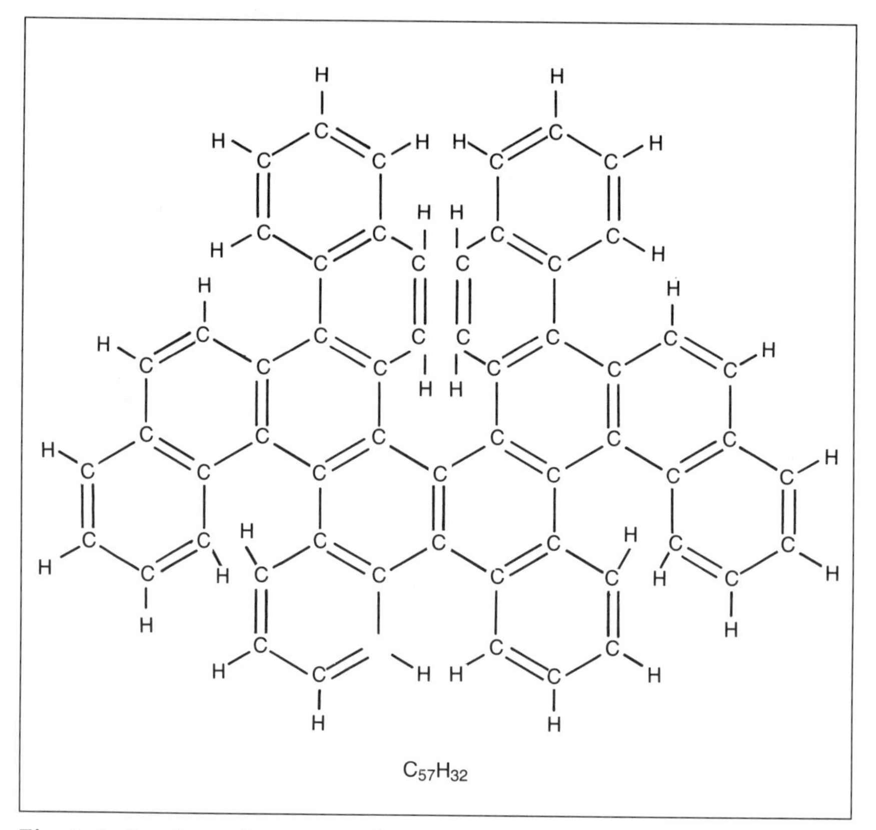

Fig. 2–1 A polynuclear aromatic

BENZENE IN PETROLEUM REFINING

The rapid increase in demand for benzene made obvious to oil refiners the advantages and disadvantages of petroleum as a supply source. Refiners were always looking for higher valued products; an all liquids system was more economical than the mechanical coal processing system; and many chemical companies were affiliates of oil refiners. At the same time there was a limited amount of benzene naturally available in oil. It was the development of sophisticated refining processes for increasing yields of gasoline that increased benzene availability. The new processes created benzene out of other molecules, permitting them to be recovered along with the free benzene naturally found in crude oil.

The benzene content of crude oil that comes out of the ground is typically only about 0.5–1.0%. Generally that's not enough to justify the equipment necessary to extract the benzene from the crude oil. The more important and commercial source of benzene is the catalytic reforming process which accounts for about 50% of the U.S. production. The original and still primary purpose of this process was to make high quality gasoline components out of low octane naphtha by reforming the molecules with the help of a catalyst. To be specific, the feed to a catalytic reformer is a mixture of paraffins and cyclic compounds in the C_6 to C_9 range. This mixture is generally called naphtha. Typically a catalytic (cat) reformer changes the naphtha composition. In the process, as illustrated in Figure 2–2,

- paraffins are converted to iso-paraffins;
- paraffins are converted to naphthenes; and
- naphthenes are converted to aromatics, including benzene.

These are the good things that happen in the cat reforming process because iso-paraffins, naphthenes, and aromatics each have higher octanes than the molecules from which they were created. Other changes happen that are not so good.

- paraffins and naphthenes can crack to form butane and lighter gases.
- some of the side chains (usually methyl groups) attached to the naphthenes and aromatics can break off also to form butanes and lighter gases.

The results of both are lower octane and economic compounds than before.

The typical change in the composition of naphtha as it passes through the reformer is shown in Table 2–1.

Table 2–1 Composition change in a cat reformer

	Volume Percent	
	Feed	Product
Paraffins	50	35
Naphthenes	40	10
Aromatics	10	55

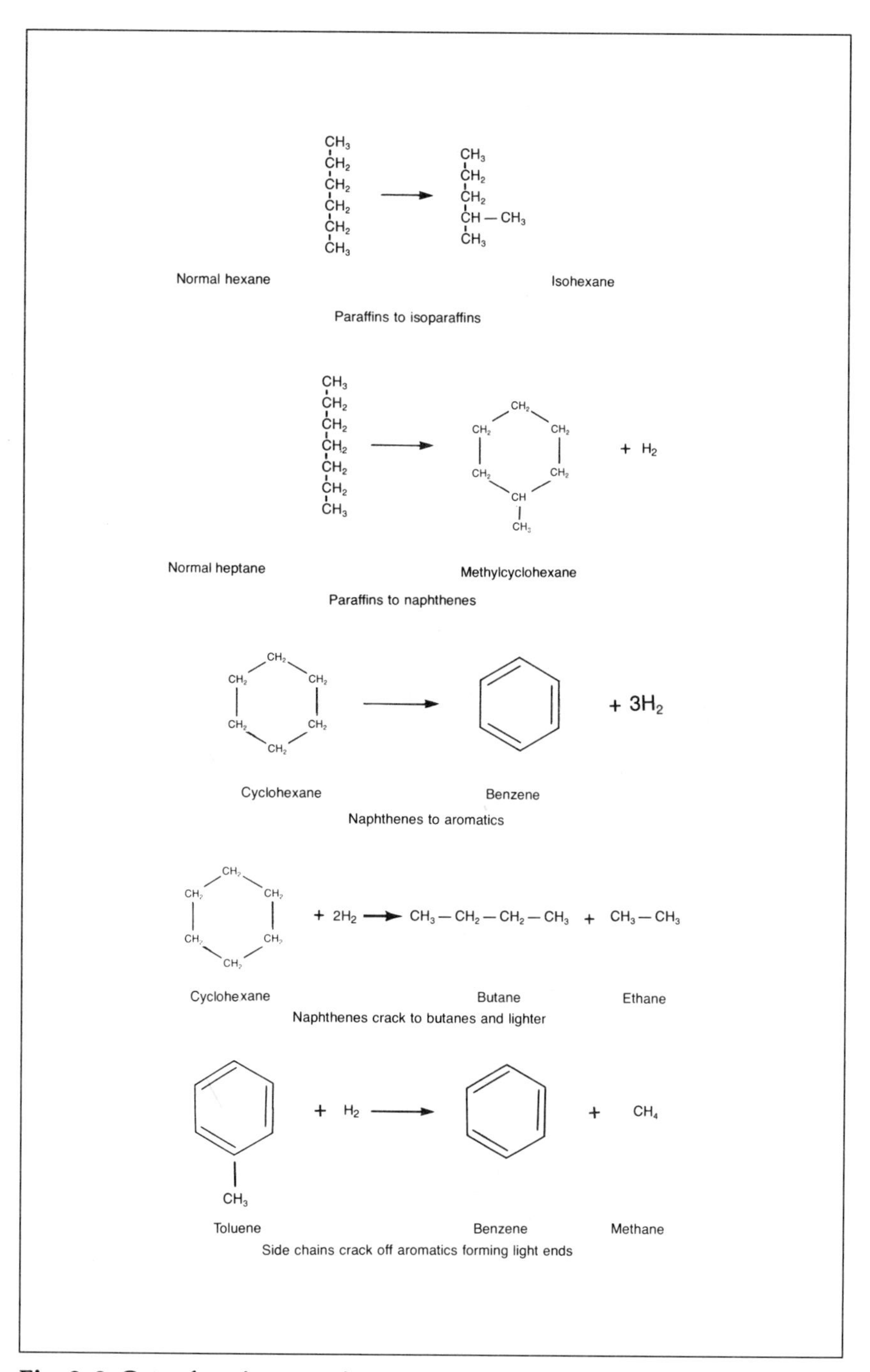

Fig. 2–2 Cat reforming reactions

The Process

These reactions are all promoted and controlled in the reactors illustrated in Figure 2–3. The naphtha is pumped through continuously at high temperatures (850–950°F) and pressures (200–800 psi). But even these severe conditions won't do it. The key ingredient is the presence of a catalyst. Each reactor is packed with pellets made of alumina or silica and coated with platinum, the catalyst. As the naphtha comes in contact with the platinum, various reactions take place, depending on the temperature and pressure in that particular reactor. Generally, there are several reactors so different sets of operating conditions can be handled, each one aimed at promoting one of the desirable reactions listed above. The platinum catalyst, by the way, does not take part in the chemical reactions. It just promotes them. Catalysts are a lot like some 10 year old kids you know. They never get into trouble. It's just that wherever they go, trouble happens.

From the chemical equations in Figures 2–2, you can see that some of the reactions give off hydrogen, while others use it up. For this reason, when the product comes out of the last stage, hydrogen is separated and recycled to be mixed with the incoming feed. This provides an abundant supply during the reactions, which is necessary to prevent the formation of some small amounts of coke during the cracking reactions. The coke will deposit itself on the catalyst, causing it to deactivate. The presence of excess hydrogen causes most of the coke to unite with the hydrogen to form light paraffins (mostly methane and ethane).

Eventually the catalyst becomes deactivated anyway, and the reactor must be shut down and regenerated. Otherwise the amount of conversion of feed to the desired products declines rapidly. Regeneration is done primarily by pumping very hot air through the reactor. The oxygen in the air reacts with the carbon on the catalyst forming carbon dioxide, which is then just blown into the atmosphere. Eventually, after a lot of deactivation and regeneration, the catalyst starts to collapse or become contaminated with other elements and must be replaced. Spent catalyst still contains all the original platinum, so it has a very high salvage value.

The amount of benzene that gets produced in a reformer will depend on the composition of the feed. Every crude oil has naphtha with

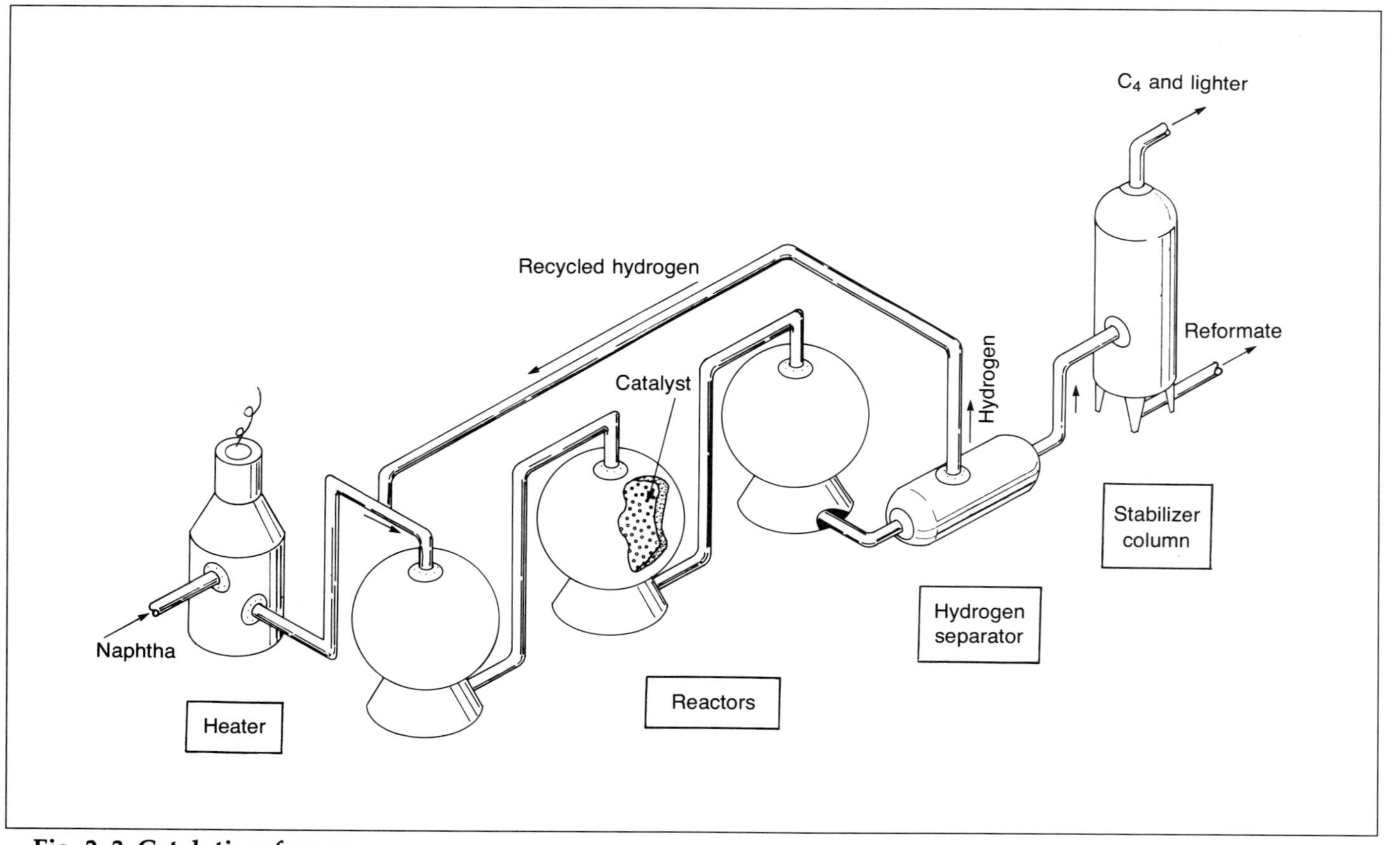

Fig. 2–3 Catalytic reformer

different PNA (paraffin, naphthene, aromatics) content. In commercial trading of naphtha, the PNA content is often an important specification. High naphthene and aromatic content would indicate a good reformer feed. High paraffin content would indicate a good olefin plant feed.

The benzene yield also will depend on the mode in which the reformer is run. For example, setting the operating conditions to maximize benzene production, will generally mean a sharp increase in the production of light ends—butanes and lighter gases. That's okay if you're not concerned about the loss of the other components, the ones used for gasoline. But if you are trying to maximize gasoline volume, benzene outturn may suffer.

The yields from a reformer, then, are a function of the feed composition, the operating conditions which are in turn responsive to economic incentives.

Downstream of the reactors and the hydrogen separator, the product is fed to one or more fractionating columns, where it is split into several streams. If just the butanes and lighter gases are removed, the remaining stream is generally called reformate. But in those refineries where benzene and/or aromatics are recovered, the reformate has removed from it a "heart cut" which has concentrated in it all the benzene in a narrow boiling range fraction. Often it is called, reasonably enough, benzene concentrate or aromatics concentrate. Benzene concentrate is about 50% benzene, plus some other C_5's, C_6's, and C_7's, all of which boil at about 176°F, the boiling point of benzene. Since the boiling temperature of the benzene is so close to that of the other hydrocarbons in the concentrate stream, fractionation is not a very effective way of isolating the benzene. Instead, a solvent recovery process called a benzene extraction unit or an aromatics recovery unit is used.

Solvent Extraction

There are certain compounds that have the remarkable characteristic of being able to selectively dissolve some compounds and at the same time ignore others. A familiar example might be to take a spoonful of table salt and drop it into a half a glass of paint thinner. The salt sinks to the bottom of the glass. Mix it, shake it, it still settles down to the bottom because it won't dissolve in paint thinner. Having observed that, it would be tough to get that salt completely separated from the paint thinner.

Now add a half a glass of water and stir. The salt disappears as it dissolves in the water. Now all you have to do is to separate the paint thinner and water by carefully pouring off the paint thinner, which is floating on the top. Then you just need to let the water stand for a couple of days and in the bottom of the glass you've got nearly all the salt you started with.

In petrochemical language, in this example:

salt-laden paint thinner is a concentrate,
water is the solvent, and
salt is the extract

Solvent extraction of benzene works the same way. But instead of water, the various solvents used are sulfolane, liquid SO_2, diethylene glycol, and NMP (N-methyl pyrrolidone). The paint thinner/salt/water process described above might be called a batch solvent process, since it consists of sequential steps which can be repeated, batch after batch. Some commercial extraction processes still operate that way. A batch of benzene concentrate is mixed with a solvent; the benzene dissolves in the solvent; the solvent separates from the undissolved hydrocarbons; the benzene-laden solvent is then drawn off and fractionated. (This step is easy by design, because a solvent is chosen that has a boiling temperature much different from the benzene.) The fractionation products are solvent and benzene.

Knowing how the batch process works, you'll find the continuous flow process just as simple. In Figure 2–4 the benzene concentrate is pumped into the bottom of a vessel with a labyrinth of mixers inside. Sometimes the mixers are mechanically moved to achieve better extraction effectiveness. A rotating disc contactor is illustrated in Figure 2–4. The solvent is pumped in the top. Almost all the solvent works its way to the bottom; the benzene concentrate works its way to the top. As the two slosh past each other, the benzene is extracted from the concentrate, passing into the solvent.

The benzene-laden solvent is handled just like the batch process—it is fractionated to separate the benzene from the solvent; the solvent is recycled back to the mixing vessel.

The remnant hydrocarbons that are taken from the top of the mixing vessel are often called benzene raffinate, a misleading, ironic name. Benzene raffinate contains no benzene. It is the leftovers after the goodies are removed. But it is still a good gasoline blending component.

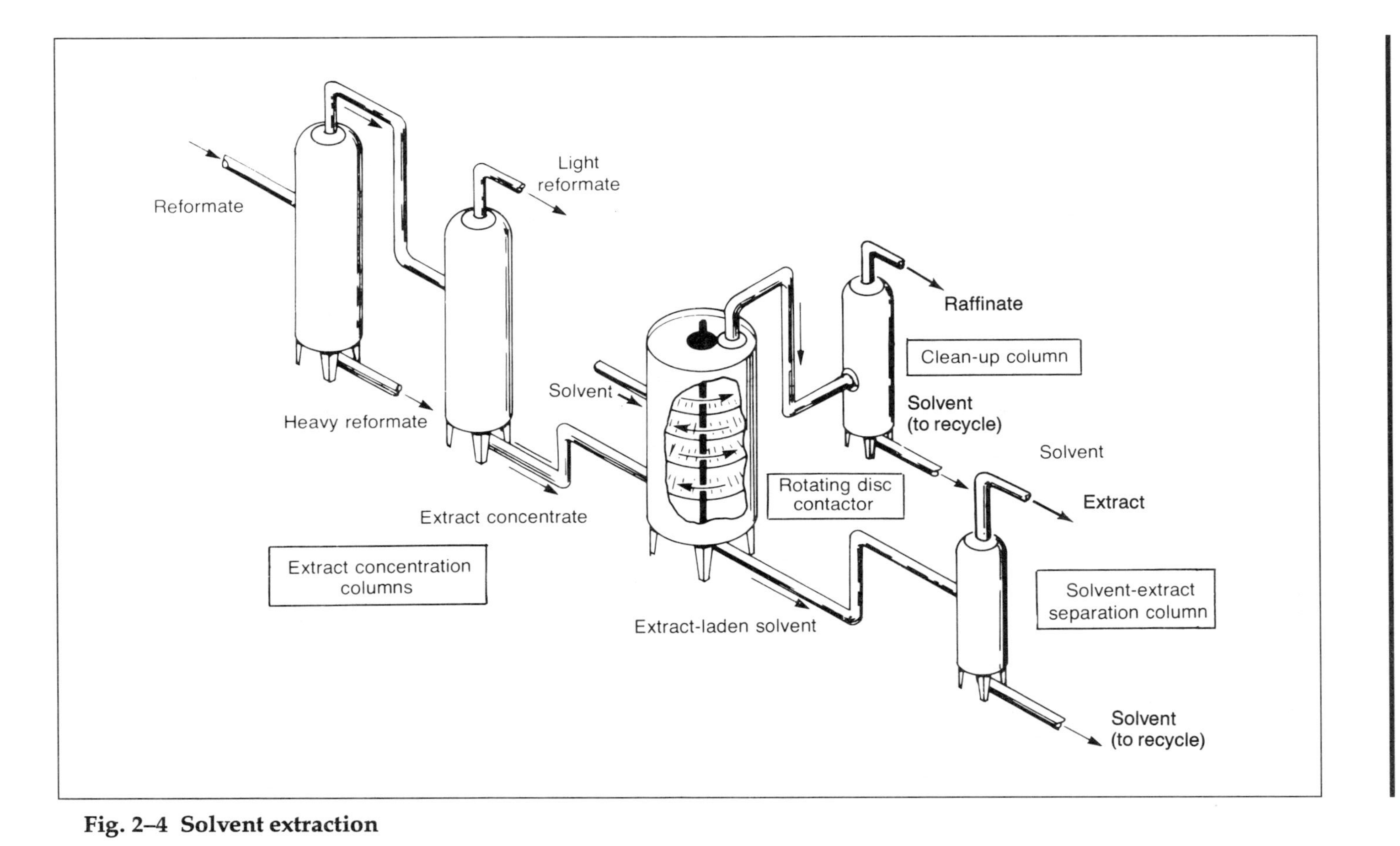

Fig. 2–4 Solvent extraction

BENZENE FROM OLEFIN PLANTS

In Chapter 4 you'll find a complete discussion of the manufacture of ethylene and propylene by cracking naphtha or gas oil in an olefin plant. One of the by-products of cracking those feedstocks is benzene. The term "by-product" may not be appropriate any more, since over 20% of the benzene supply in the U.S. now comes from olefins plants.

Naphthas and gasoils consist of molecules with carbon counts of 5 to 20 or more. The olefins are created by heating the molecules to a temperature where they crack, forming among other things the desired ethylene (C_2H_4) and propylene (C_3H_6). The larger carbon count molecules, C_{10} and higher, often contain multiple benzene rings, not too unlike the coal configuration described above. When the molecules break up, the benzene rings can be freed intact, forming benzene and other aromatics. The process is similar to the destructive distillation of coal, when it comes to benzene.

The benzene leaves the olefins plant mixed with the other gasoline components. Recovery is done in the same manner already described, in a benzene recovery unit.

BENZENE FROM TOLUENE HYDRODEALKYLATION

Since toluene is nothing more than benzene with a methyl group attached, creating one from another is relatively easy. Benzene, toluene, and for that matter, xylenes too, are co-produced in the processes described above—coke making, reforming, and olefin plants operations. The ratio of production is rarely equal to the chemical feedstock requirements for the three. One method for balancing supply and demand is toluene hydrodealkylation (HDA). This process accounts for about 30% of the supply of benzene in the U.S. and is a good example of what can be done when one or more co-products are produced in proportions out of balance with the marketplace.

The word hydrodealkylation is less ominous than it appears. Alkanes are a synonym for paraffins; alkylation is the process of adding a paraffin group (like a methyl or ethyl group) to another compound. Dealkylation is nothing more than removing it. Hydro- indicates the replacement atom is hydrogen.

In the toluene HDA process shown in Figure 2–5, toluene is mixed with a hydrogen stream, heated, and pumped into a reactor. This vessel, like a cat reformer reactor, is packed with a platinum catalyst and runs at high pressures and temperatures. The methyl group pops right off as the toluene passes over the catalyst. Hydrogen fills out the valence requirements of the resulting molecule, forming benzene.

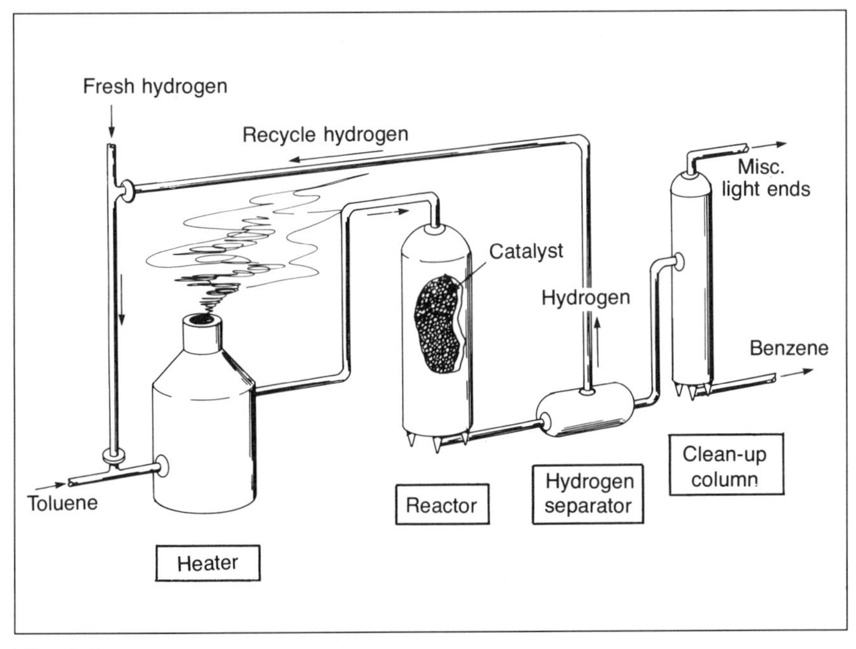

Fig. 2–5

The stream leaving the reactor is separated in several fractionators into hydrogen, methane and other light gases, and benzene. The hydrogen is recycled, and the light gases are usually sent off to the fuel system. The benzene can be clay treated to remove any stray olefins, resulting in a pure, nitration grade benzene.

The yield of benzene in a toluene HDA plant runs 96–98%.

Material Balance	
Feed:	
Toluene	1200 lbs.
Hydrogen	27 lbs.
Product:	
Benzene	1000 lbs.
Methane and other misc.	227 lbs.

HANDLING BENZENE

Benzene is a clear, colorless, flammable liquid with a distinct, sweet odor. It burns with a smokey flame, as do other hydrocarbons with high aromatic content. (That's why kerosenes with high aromatic content do not make good jet fuel or burning grade kerosene—too much black smoke.) Benzene is only slightly soluble in water.

Freezing point	41.9°F (5.5°C)
Boiling point	176.2°F (80.1°C)
Specific gravity	0.879 (lighter than water)
Weight per gallon	7.32 lbs/gal

The commercially traded grades of benzene are motor or industrial, industrial pure (2°F boiling range), and nitration grade (1°F boiling range). The boiling range is a measure of the amount of impurities (other hydrocarbons) mixed in with the benzene. The wider the boiling range, the more impurities. Industrial pure benzene has about 0.5%; nitration grade has even less because it is clay filtered to remove the more reactive compounds like thiophene, a sulfur-containing, bad-smelling heterocyclic. Motor benzene is generally mixed into gasoline, so it can stand the boiling range of 7°F.

Benzene is shipped in tank cars, tank trucks, barges, and drums. Transfers from one vessel to another are in closed systems because benzene is a poisonous substance with acute toxic effects. It'll kill you in 5–10 minutes if you breath too much. Red DOT flammable liquid labels are required.

USE PATTERNS

Most of the benzene used in chemical applications ends up in the manufacturing processes for styrene (covered in Chapter 8), cumene (covered in Chapter 7), and cyclohexane (covered in Chapter 4). Polymers and all sorts of plastics are produced from styrene. Cumene is the precursor to phenol, which ultimately ends up in resins and adhesives, mostly for gluing plywood together. The production of styrene and phenol account for about 70% of the benzene produced. Cyclohexane, used to make Nylon 6 and Nylon 66, is the next biggest application of benzene.

Other smaller but important volumes of benzene end up in the processes for making maleic anhydride (for resins), nitrobenzene (for explosives), aniline (for dyes), and dodecylbenzene (for detergents).

Chapter II in a nutshell...

Benzene, C_6H_6, is a ring of carbon atoms, connected alternately by a single and a double bond. Each carbon has a single hydrogen attached. It is found as a natural component in crude oil; it is created in the process of catalytically reforming naphtha to make high octane gasoline components; and it is formed in thermal cracking processes such as an olefins plant where complex molecules containing benzene rings are split up. High purity benzene is extracted in a special solvent recovery unit. Benzene is used in the production of numerous chemicals including styrene, cumene, cyclohexane, and maleic anhydride.

Exercises

1. Assemble the following list into groups of feeds, operating units, and outturns:

benzene	coal	naphtha
benzene	coke	olefin plant
benzene	destructive distillation	solvent extraction unit
benzene	gas oil	toluene
cat reformer	hydrodealkation	reformate

2. If you had 500,000 gallons of toluene and a toluene HDA unit and the toluene market price was \$0.20/lb., benzene was \$0.24/lb., hydrogen was \$0.40/lb., and it cost \$0.005/lb. to run the HDA unit, what would you do? Oh, and toluene weighs 7.21 lbs. per gallon and benzene is 7.32 lbs/gallon.
3. What is the "cat" in cat reformer?
4. What's the difference between reformate and raffinate?
5. Some coffee companies use methylene chloride to take the caffeine out of regular coffee. In this solvent extraction process, what do you think are the solvent, the raffinate, the extract, and the feed?

TOLUENE AND THE XYLENES

"Into fire,
into ice."

■

Divine Comedy
Dante, 1265-1321

Should this be a separate chapter? The chemistry and hardware involved in making toluene and xylenes are for the most part the same as their sibling, benzene. While that may be true, there are a few chemical principles that can be demonstrated better using toluene. The separation processes for purifying toluene and xylenes are different also. There's enough, then, for a healthy bite without tagging on to the last chapter.

TOLUENE

The manufacture routes to toluene, like benzene, are coke production, cat reforming, olefin plant operations, and recovery of the small amount naturally occurring in crude oil. And also like benzene, today most toluene comes from cat reforming. Coal-derived toluene became a minor share of the market in the 1950s.

In the cat reforming process, there are two importantly controllable variables: the composition of the feed and the operating conditions in the reactors. As to the first, some compounds are more suitable for reforming into toluene than others. These precursors (from the Latin *curro*, I run, and *pre*, before) or forerunners include cyclohexane, methyl cyclohexane, ethyl cyclopentane, and dimethyl cyclopentane. In Figure 3–1, you will notice that three of these compounds have the same carbon count, C_7H_{14}, and the carbon count is the same as toluene, C_7H_8. Three different types of reactions take place in a cat reformer that change the precursers to toluene: ring opening, dehydrogenation, and cyclicization. Just looking at Figure 3–1 you can imagine that dehydrogenation (the removal of hydrogen) is necessary to work on the methyl cyclohexane. Because ethyl cyclopentane and dimethyl cyclopentane start out with the wrong carbon number in their rings, both ring opening and cyclization (closing a ring back up again), as well as dehydrogenation, are needed to get to toluene.

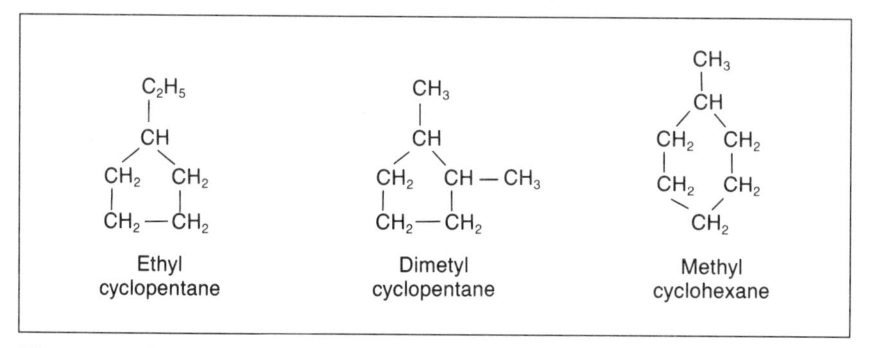

Fig. 3–1 Toluene precursors

When naphthas with a high natural content of these precursers are used in cat reformers, the yields of toluene are high. Unfortunately (for the toluene merchant) there's generally not too much attention paid to toluene manufacture optimization in refineries, for several reasons:

a. more toluene generally means less benzene.
b. the composition of the naphtha feed depends on the selection of the crude oil, and that is usually determined by factors other than

reformer operations because the reformer feedstock fraction is only a small part of crude oil.

c. Most reformate, which contains the toluene and xylene, ends up as gasoline blending components anyway. Only a small portion of the toluene is isolated, about 15-20%. That leaves a large pool for any further expansion of its applications—almost as easy as turning on that spigot.

Separation of toluene from the other components can be by solvent extraction, together with the benzene, or in separate streams. The boiling points of benzene and toluene are far enough apart that the feed to the solvent extraction can be split (fractionated) rather easily into a benzene concentrate and a toluene concentrate. Alternatively, an aromatics concentrate stream can be fed to the solvent extraction unit, and the aromatics outturn can be split into benzene and toluene streams by fractionation. Both schemes are popular.

Azeotropic Distillation of Toluene

There is an alternate process for recovering toluene from the reformate stream called azeotropic distillation. It also can be used to split toluene from the other hydrocarbons that have boiling points near toluene. It is like solvent extraction with an extra twist. But the process can be more efficient than extraction when the toluene concentration is high.

In azeotropic distillation, a solvent is used which increases the volatility of the components to be removed. In this case what is removed is everything in the toluene concentrate but the toluene. The added solvent, along with the unwanted components go up the distilling column as a vapor; the toluene goes down and out as a liquid.

An example might help. An occasional automotive problem is water in the gasoline. It's usually caused by warm moist air getting in a half empty gas tank, and then the water condensing when the weather turns cold. Water in gasoline causes hard starts and sputtering because it won't vaporize in the carburetor. An over-the-counter remedy is dry gas, which is nothing more than ethyl alcohol. Water will dissolve in the alcohol, and together they will act just like gasoline as they go through the carburetor because together they vaporize at a lower temperature than either water or ethyl alcohol alone. In this analogy, the solvent is ethyl alcohol, the (toluene) extract is gasoline, and the raffinate is water.

When azeotropic distillation is used for toluene, the solvent used is usually a mixture of methyl ethyl ketone (MEK) and water (10%). The solvent and the toluene are mixed, heated, and then charged to a distillation column (Fig. 3–2). The paraffins and naphthenes dissolve in the MEK/water and then vaporize about 20 °F lower than normal. The vapors work their way up the distilling column; the toluene works its way down as a liquid. Again, this takes place despite the fact that the paraffins and the naphthenes have nearly the same boiling temperatures as toluene. The solvent does it.

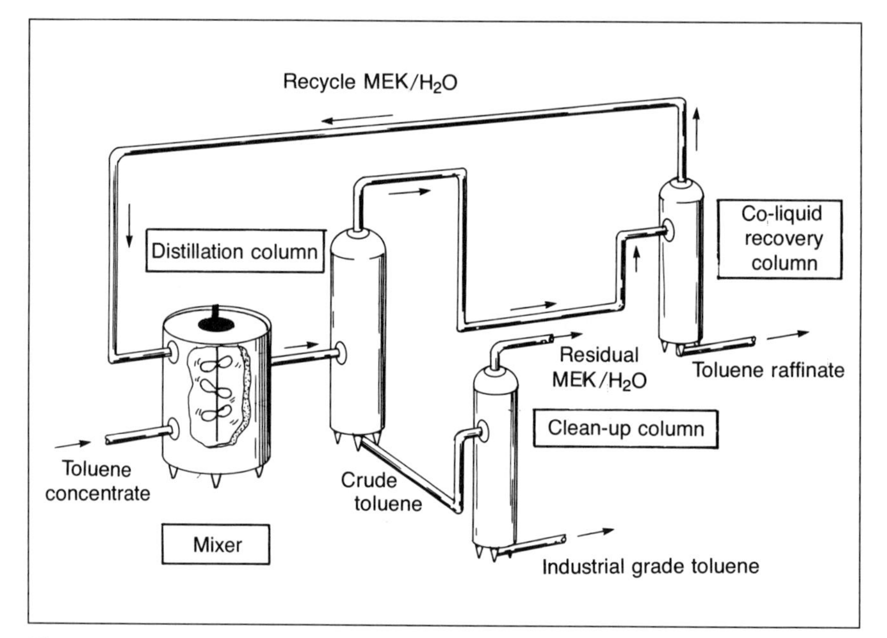

Fig. 3–2 Azeotropic distillation of toluene

Commercial Use

Toluene, like benzene, is a flammable liquid and requires the red DOT shipping label. There are two commercially available grades, usually defined in terms of boiling ranges. Industrial grade toluene (95 to 98%) boils within two degrees of the toluene boiling point of 213 °F (100.6 °C). Nitration grade toluene (99%) boils within a one degree range. The term "nitration" grade is a hangover from the specification required for the manufacture of trinitrotoluene (TNT). Lower grades are known as crude toluene, aviation grade toluene, or other specialty names.

Use Patterns of Toluene

During World War II, two militarily, but not chemically, related uses gave a running start to toluene. Because of its high octane characteristics (103–106 octane number), toluene was particularly suitable for blending aviation gasoline. Wartime conditions made maximum production of toluene an imperative during this period.

At the same time, the need to manufacture military explosives caused the demand for toluene for TNT to peak. Ironically, the chemistry that makes for good octane characteristics has little to do with that of explosives.

In the postwar period, the expansion of commercial aviation sustained the demand growth for toluene as a high octane blending component. By the 1960s, aviation gasoline gave way to jet fuel in most commercial aircraft. But the growth of automotive gasoline and the accompanying octane wars more than compensated, and today gasoline remains the largest application for toluene.

Yet toluene remains an important petrochemical building block. The major use is in hydrodealkylation units, making benzene. Over 50% of the recovered toluene is used in this way in the U.S. Conversion to para-xylene is also of growing importance.

Toluene is used more commonly as a commercial solvent than the other BTX's. There are scores of solvent applications, though environmental and health concerns are diminishing the enthusiasm for these uses. Toluene also is used to make toluene diisocyanate, the precurser to polyurethane foams. Other derivatives include phenol, benzyl alcohol, and benzoic acid. Research continues on ways to use toluene in applications that now require benzene. The hope is that the dealkylation-to-benzene step can be eliminated. Processes for manufacturing styrene and for terephhtalic acid—the precurser to polyester fiber—are good, commercial prospects.

XYLENE

You might think that to finish out the aromatics family, the X in BTX would be the xylene triplets, ortho-, para-, and meta-xylene. But there's another isomer in the closet, ethylbenzene. It has the xylene carbon and hydrogen count, C_8H_{10}, but it's a benzene ring with an ethyl group (-C_2H_5) attached, not two methyl groups (-CH_3). Ethylbenzene will figure importantly in the chapter on styrene.

The manufacture of the xylenes is similar to benzene and toluene—cat reforming, olefin plants, coke making, plus a small amount naturally resident in crude oil. Minor amounts are made by catalytic disproportionation, a process in which the methyl group is clipped off one toluene molecule (forming benzene) and ends up on another (forming xylene.)

What really makes the xylenes different from the other BTX's are the techniques to separate them from each other. That will be the main topic addressed in this section.

Process

The BTX's generally come from two sources but are handled in the same processing scheme (Fig. 3–3). In a refinery the reformate stream coming from the cat reformer contains most of the aromatics, together with miscellaneous naphthenes (alicyclics) and paraffins (aliphatics). The BTX's in the stream coming from an olefins plant also has some olefins mixed in as well. The BTX's are recovered as a mixture in the solvent extraction unit. The boiling points of benzene, toluene, ethyl benzene, and ortho-xylene are far enough apart that they can be separated by simple fractional distillation. Meta-and para-xylene are not. But take a look at the freezing points for these two:

	Boiling Points °F	Freezing Points °F
Benzene	176.2	41.9
Toluene	231.4	−138.9
Ortho-xylene	292.0	−13.0
Meta-xylene	282.4	−54.2
Para-xylene	281.0	55.9
Ethylbenzene	277.1	−138.0

Ortho-xylene can be separated from others by distillation; ethylbenzene is only 3.9°F from para-xylene, but using very tall, multi-trayed distillation columns (200 feet high with 300 trays), those too can be separated fairly completely. But the 1.4°F spread between meta-and para-xylene requires columns more expensive than chemical companies could stand, so alternate separation techniques were developed: cryogenic crystallization and adsorption, using molecular sieves.

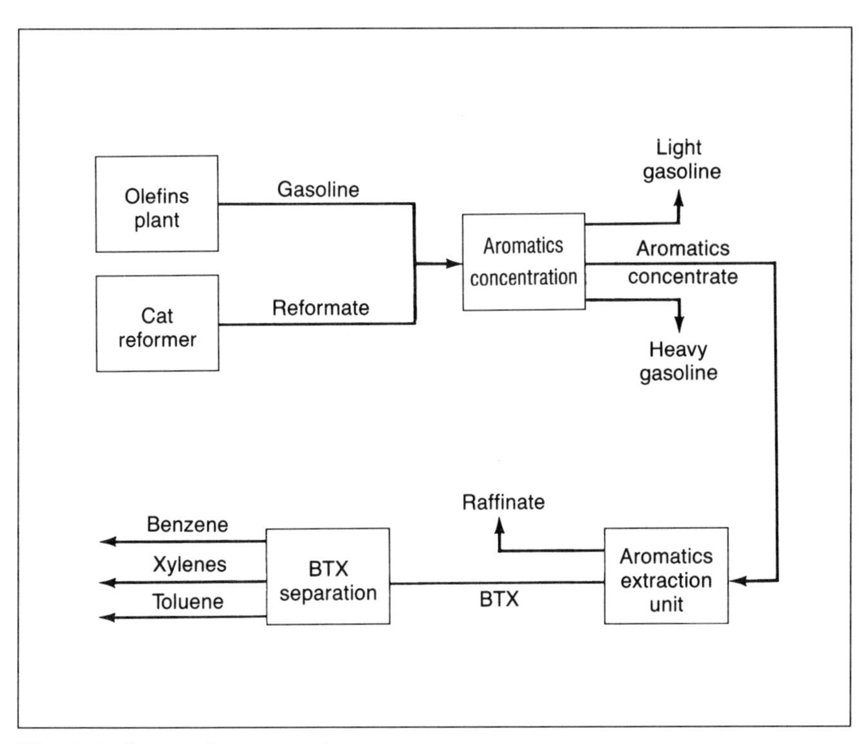

Fig. 3–3 Aromatic processing

Cryogenic Crystallization. Even though the boiling temperatures of meta-and para-xylene are close together, their freezing points, i.e., the temperatures at which the liquid starts freezing, i.e., turning to crystals, are not. Meta-xylene crystallizes at -54.2°F and para-xylene at +55.9°F.

In Figure 3–4, the processing scheme shows the ortho-xylene and ethylbenzene split out in fractionators. The mixed para- and meta-xylenes are then processed in a fashion a lot like making good pot roast gravy. In order to keep the grease out of the gravy and if you've got the time, you can put the beef drippings in the refrigerator for an hour or two. All the grease solidifies and floats to the top, and can be spooned off and discarded. Similarly, the mixed para- and meta-xylenes are cooled initially to about -90°F in a holding tank. At that temperature, para-xylene crystals form and grow, forming a liquid-solid mixture like slush. The key to good solid-liquid separation is large crystal growth. The larger the crystals, the better the separation because of the next step.

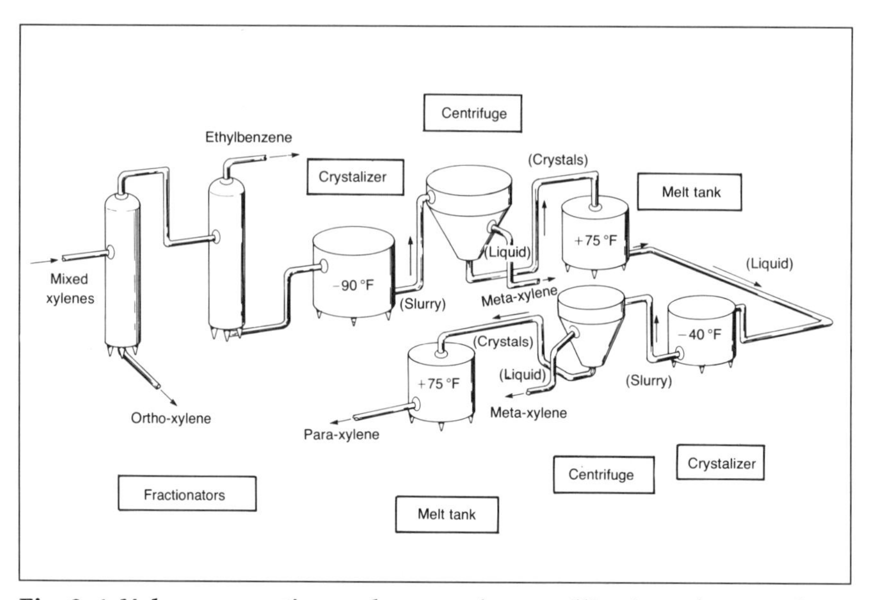

Fig. 3–4 Xylene separation and cryogenic crystallization of para-xylene

When the crystals have grown sufficiently, the slush is put in a centrifuge. The spinning action permits the para-xylene to separate from the mother liquor, so-called because the crystals come out of the liquid. At this stage, the filter cake (the para-xylene crystals) has a purity of 80 to 90%, due to the mother liquor which coats the crystal surface. (That's the reason for big crystals—less surface area for the mother liquor to coat.)

To further purify the para-xylene, the crystals are again melted, cooled—this time to about -40°F—and crystallized once again. Centrifuging this time results in about 99% pure para-xylene. The meta-xylene from both centrifuges is about 85%.

Adsorption Process. A commercial process was introduced in the 1970s using a material called molecular sieve, which can separate the para-xylene from the meta-xylene. Molecular sieves are marble-sized pellets that have millions of pores, all of a size that para-xylene molecule can fit in but the meta-xylene molecule cannot. The pore sizes are so small they are measured in Angstroms, which are $1x10^{-8}$ centimeters (0.00000001 cm). Molecular sieves of varying pore sizes are used in many other applications as well.

In an adsorption process, molecules are collected on the surface of a substance or agent and held there by electrostatic force. That is in contrast to absorbtion, where the collection takes place inside the agent. Hence, the uncommon suffix, ad-. And, when the adsoption agent is a molecular sieve, it might seem a little confusing because the collection takes place on the surface of the pores, which might seem more like ab- than ad- , but isn't.

In a para-xylene plant, a column or bed packed with molecular sieve pellets is used. The mixed para- and meta-xylenes are pumped through the bed. At first, the liquid stream coming out of the bed (the effluent stream) is very low in para-xylenes, because they are selectively being collected in the sieves. Gradually the concentration of para-xylene in the effluent stream starts to build as the molecular sieve pores fill up and the para-xylene slips by. At some point, the operation is shut down. A fluid is then pumped backwards through the bed to flush out the para-xylene. The process of pulling the para-xylene from the molecular sieve is called desorption, in which a fluid called a desorbent is used to flush the para-xylene out. The desorbent used is chosen so that its separation from the para-xylene in a distillation column is easy.

After the desorption, the bed is heated up to vaporize and remove all the desorbent and remaining para-xylene. The cycle is then ready to begin again. Para-xylene purity from this technique is about 99.5%. The adsorption process presently has some economic advantage over the cryogenic crystallization route, due mainly to fuel and operation costs.

Commercial Use. The composition of mixed xylenes depends on the cat reformer operations at a refinery and the type of crude oil and naphtha being processed. Generally the yield of mixed xylenes and ethylbenzene is greater than the combined yield of benzene and toluene. Typical composition would be:

Ethylbenzene	10 to 15%
Para-xylene	17 to 20%
Meta-xylene	46 to 52%
Ortho-xylene	18 to 24%

Mixed xylenes are commercially available in nitration grades which have tolerances of 3 and 10°F, depending on the specified amount of the hydrocarbon present. Purities of the ortho-, meta- and para-xylenes are more often than not a matter of negotiation between buyer and seller. Ethylbenzene as a separate stream is generally not a commercial commodity. Its primary use in petrochemicals is as an intermediate stream on the way to making styrene from benzene and ethylene.

Mixed xylenes are used as an octane improver in gasoline and for commercial solvents, particularly in industrial cleaning operations. By far, most of the commercial activity is with the individual isomers. Para-xylene, the most important, is principally used in the manufacture of terephthalic acid and dimethyl terephthalate enroute to polyester plastics and fibers (Dacron, films such as Mylar, and fabricated products such as plastic bottles). Ortho-xylene is used to make phthalic anhydride, which in turn is used to make polyester and alkyl resins and PVC plasticizers. Meta-xylene is used to a limited extent to make isophthalic acid, a monomer used in making thermally stable polyimide, polyester, and alkyd resins.

There is usually more meta-xylene around than needed for its limited derivatives, so some of it is processed in isomerization units and converted to ortho- and para-xylene.

The xylenes are flammable and are shipped under the same regulations and using the same methods as benzene and toluene: tank cars' trucks, barges, and tankers. Pipeline movements are limited. Toxicological problems dictate handling in closed systems like benzene and toluene.

Chapter III in a nutshell...

Toluene, $C_6H_5CH_3$, and the xylenes, $C_6H_4(CH_3)_2$, are benzene rings with one or two methyl groups, $-CH_3$, hung on in the place of hydrogens. Toluene has one; the xylenes have two. Sources of all three BTX's are the same: crude oil, catalytic reforming, heavy liquids cracking in an olefins plant, and to a declining extent, coking at a steel plant.

BTX's are recovered in solvent extraction plants. Toluene and the xylenes are separated by unique methods. Toluene is usually separated by azeotropic distillation using co-solvents. Para-xylene separation is usually done by adsorption using molecular sieves or by cryogenic crystallization.

All the BTX's are high octane gasoline blending components. In the petrochemicals business, toluene is used as a building block for polyurethane. Para-xylene and ortho-xylene are used to make polyester fibers and plastics, alkyd resins, and plasticizers.

Exercises

1. Name the six compounds that make up the BTX's.
2. What goes in and what comes out of the following plants?

feeds	plants	product
toluene	adsorption	benzene
toluene	cryogenic distillation	mixed xylenes
mixed-xylenes	disproportionation	ortho-xylene
mixed-xylenes	hydodealkylation	para-xylene

3. Can a mixture of benzene, toluene, and meta-xylene be separated by cryogenic crystallization? What's the usual (more economic) way?
4. Ortho-, para-, and meta-xylene are:
   a. isomers
   b. aromatics
   c. solvents
   d. petrochemical feedstocks
   e. all of the above
   f. the three ugly daughters of a mad Hungarian named Vladok Xylene.

IV

CYCLOHEXANE

"A hen is an
egg's way of making
another egg."

■

Life and Habit
Samuel Butler, 1835–1902

The petrochemical business is funny. Some companies use cyclohexane to make benzene. Some use benzene to make cyclohexane. This chapter covers the latter.

It was the development of nylon by DuPont in 1938 that charged up the interest in cyclohexane. They settled on the use of cyclohexane as their preferred raw material. In the period right after World War II, the manufacture of nylon grew for a while at 100% annually, quickly overwhelming the availability of cyclohexane naturally available in crude oil. The typical crude oil processed in U.S. refineries at the time had less than 1% content of cyclohexane. Ironically, since cyclohexane leaves the crude oil distillation operation in the naphtha, it was usually fed to a cat reformer, where it was converted to benzene. As it turned out, with so many other precursers also being converted to benzene in the cat reformer, benzene became a good source for cyclohexane.

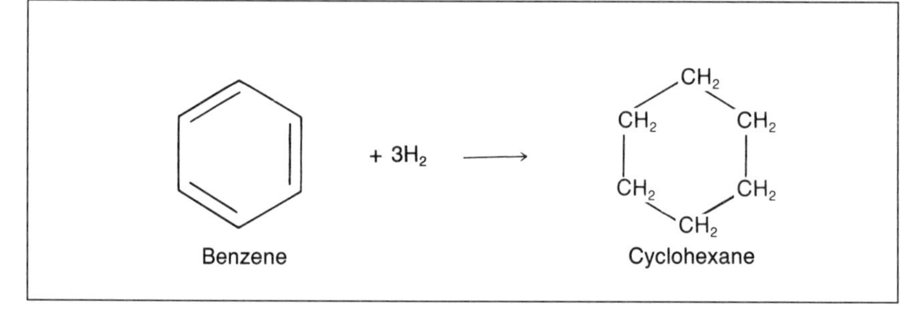

Fig. 4–1 Benzene hydrogenation to cyclohexane

As a source of cyclohexane, benzene has the right shape (Figure 4-1) but too few hydrogens. So cyclohexane plants are not much more than vessels in which benzene molecules are hydrogenated with the help of a catalyst. This process accounts for about 90% of today's cyclohexane.

Benzene, you'll recall, has alternating double bonds, and the addition of one hydrogen atom to any one of the carbons will cascade quickly all around the benzene ring so that all the carbons pick up a hydrogen. Pressure and temperature alone cannot cause the hydrogenation—a catalyst is needed. Fortunately, several metals qualify—platinum, palladium, nickel, and chromium. The first two are highly active, and can cause hydrogenation to occur at room temperature and only 15 to 20 psi pressure. Unfortunately, platinum and palladium are expensive metals. Most commercial processes use nickel or chromium. Though they require much higher temperature and pressures, which is more expensive in terms of energy costs, the catalyst is cheaper.

Sulfur and carbon monoxide can be killers (literally) with hydrogenation catalysts. It will "poison" them, making them completely ineffective. Some sulfur often shows up in the benzene feed, carbon monoxide in the hydrogen feed. The alternatives to protect the catalyst are either to pre-treat the feed and/or the hydrogen or to use a sulfur resistant catalyst metal like tin, titanium, or molybdenum. The economic trade-offs are additional processing facilities and operating costs versus catalyst expense, activity, and replacement frequency. The downtime consequences of catalyst replacement usually warrant the more expensive front-end treatment facilities.

THE PROCESS

The hardware used for the hydrogenation of benzene is shown in Figure 4–2. The basic parts are three or four reactors in a series plus a separation section at the end. The reactors are vessels filled with catalyst in the form of charcoal or alumina pellets that are coated with one of the above metals. The catalyst is packed loosely enough that the feed can flow through, top to bottom.

The continuous flow process shown in Figure 4–2 has a mixture of benzene, cyclohexane, and hydrogen being heated to about 400 °F, pressured to about 400 psi, and pumped through the first reactor. The proportions of each feed depend on the type of catalyst being used. On a once through basis, about 95% of the benzene is converted to cyclohexane.

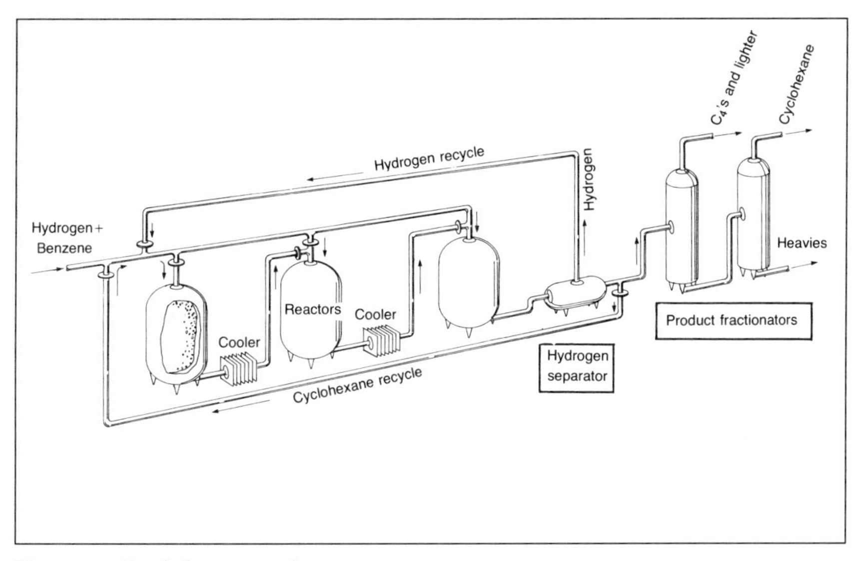

Fig. 4–2 Cyclohexane plant

Most hydrogenation reactions, this one included, are exothermic, i.e., they give off heat. To minimize the by-products that could occur, strict temperature control must be maintained. As the feed passes through the reactor like the one shown in Figure 4–3, the temperature increases by about 50 °F. The reactor effluent is therefore cooled back down to 400 °F in a heat exchanger. For the second pass, additional benzene is added, although the resulting proportion in the second and succeeding reactors keeps decreasing. The same process of hydrogenation, with its exothermic effects occurs, and the reactor effluent must be cooled again in a heat exchanger to get it to the right temperature for the next reactor.

The overall conversion of benzene to cyclohexane is nearly 100%, but the effluent from the last reactor will still have plenty of hydrogen in it. To facilitate the hydrogenation reaction, hydrogen is usually kept in excess. The effluent is passed through a flash drum, where the pressure drops and the hydrogen flashes out of the product and is recycled to the feed. The remaining effluent is then fractionated as a final cyclohexane purification step. (Since reaction conditions in the process are never controlled perfectly, some of the benzene feed, and whatever other hydrocarbons come along with it, get converted to other miscellaneous compounds, mostly butanes and lighter gases, that have to be removed in the distillation step.)

A cyclohexane stream is recycled to the feed also, and performs an important function. It acts as a heat sink or a sponge, diluting the exothermic effect of the hydrogenation reaction, keeping the temperature down. At temperatures above 450°F, the decomposition of benzene to those light ends just mentioned increases rapidly.

Material Balance	
Feed:	
Benzene	944 lbs.
Hydrogen	65
Catalyst	–
Product:	
Cyclohexane	1000 lbs.
Light ends	9

In summary, the key variables in this process are temperature control, excess hydrogen, and catalyst activity. Conversions are typically 99.5%.

COMMERCIAL ASPECTS

Cyclohexane is a colorless, water insoluble, non-corrosive liquid having a really pungent odor. It's flammable like any naphtha product, and it is shipped in tank cars, tank trucks, barges, and drums. Red DOT

Fig. 4–3 New cyclohexane plant at Lake Charles, La., recently went on stream for Continental Oil Company. Plant was designed and licensed by UOP (Universal Oil Products) for Conoco. Three conical objects in the foreground are the plant's reactors. Using benzene feedstock, plant produces very high purity grade of cyclohexane used in production of nylons. Conoco has similar plant at Ponca City, Okla.

shipping labels are required. In commerce, trade is usually done on the basis of Technical Grade (either 95 or 99% purity) or Solvent Grade (85% minimum purity).

Cyclohexane Properties	
Freezing point	43.7°F (6.5°C)
Boiling point	177.3°F (80.7°C)
Specific gravity	0.7786 (lighter than water)
Weight per gallon	6.54 lbs./gallon

Virtually all cyclohexane is used to make three intermediate chemicals: caprolactam, adipic acid, and hexamethylene diamine—the starting materials for Nylon 6 and Nylon 66 synthetic fibers and resins. Nylon fiber markets include the familiar applications: hosiery, upholstery, carpet, and tire cord. Nylon resins are engineering plastics and are largely used to manufacture gears, washers, and similar applications where economy, strength, and a surface with minimal friction are important. Minor uses of cyclohexane include industrial solvent applications such as cutting fats, oils, and rubber. It's also used in paint remover.

Chapter IV in a nutshell...

Cyclohexane, C_6H_{12}, is a carbon ring with two hydrogen atoms attached to each carbon. It resembles benzene, but there are no double bonds. Benzene is the feed to a cyclohexane plant, which is just a hydrogenation process.

Cyclohexane is a colorless liquid at room temperature. It is used primarily to make precursers of Nylon 6 and Nylon 66.

Exercises

1. What does "exothermic effect of hydrogenation reaction" mean?
2. Fill in the blanks:
   a. To make cyclohexane out of benzene, you need to add __________ atoms.
   b. Various __________ are used to promote the benzene hydrogenation reaction.
   c. If it weren't for the development of __________, most cyclohexane would still end up as a gasoline blending component or as cat reformer feed.
3. If cyclohexane is worth 30 cents/lb., hydrogen costs 40 cents/lb., and it costs 0.5 cents/lb. of feed to run a cyclohexane plant, how much can you afford to pay for benzene to break even? Assume the light ends are flared (that is, they are burned off and worth nothing).
4. Which came first, the chicken (cyclohexane) or the egg (benzene)?

V

OLEFIN PLANTS, ETHYLENE, AND PROPYLENE

"Anything that can happen
will happen."

■

Murphy's Third Law

The big daddy of the petrochemical industry is the olefin plant. The vintage of this process dates back before the 1940s. Olefin plants are a wellspring of the industry's basic building blocks—ethylene, propylene, butylenes, butadiene, and benzene. The recently built olefin plants are huge. The so-called world-scale plant (the size that achieves whatever is currently considered full economies of scale) is larger than many medium size refineries. Capacity is no longer measured in millions but billions of pounds per year!

There are a lot of options on how to design, feed, and operate an olefin plant. For that reason, this chapter will cover in some depth the hardware, the reactions, and the variables that can be manipulated to change the amount and mix of products. The physical properties of ethylene and propylene, which present some unique handling problems, will be covered also.

Olefin plants have more than one alias. (One is even fraudulent.) They are variously called ethylene plants (after their primary product); steam crackers (because the feed is usually mixed with steam before it is cracked); or ____________ cracker, where the blank space is the name of the feed (ethane cracker, gas oil cracker, etc.) Olefin plants are sometimes referred to as ethylene crackers, but that is the misnomer. Ethylene is not cracked but rather is the product of cracking.

Since ethylene is such a simple molecule, $CH_2=CH_2$, it stands to reason that there are lots of hydrocarbons that could be cracked to form it. The earliest commercial olefin plants of any size were designed to use ethane and propane. As you can see in Table 5–1, ethane and propane produced a high yield of ethylene; propane also gave a high yield of propylene. These two feedstocks were the dominant choices through the 1960s. Because of the shortages of natural gas in the 1970s, there was a perception at that time that future production of natural gas would be declining, and along with it the availability of ethane. Propane would be expensive since it would be a substitute for the disappearing natural gas. For this reason, by the late 1970s, the industry developed the technology and built the plants to crack heavier feedstocks, including naphtha and gas oils. Presently about half the ethylene comes from ethane/propane cracking; most of the rest comes from naphtha and gas oil.

The one aspect of ethylene manufacture that sets it apart from most other petrochemicals is the wide range of alternate feedstocks that can be used. Most others are limited to one or a few commercial alternatives. But many companies have their plants arranged so that they have a menu of alternatives from which to choose. So the economics of feedstocks becomes an important variable in olefin plant operations, as you will see.

THE PROCESS

Ethane and propane cracking is simpler than heavy liquid (gas oil or naphtha) cracking and should be tackled first. When ethane is heated up to 1700°F or above, either of two basic reactions can occur, splitting of

Table 5–1 Olefin Plant Yields

	Pounds Per Pound of Feed				
	Ethane	Propane	Butane	Naphtha	Gas Oil
Yield:					
Ethylene	0.80	0.40	0.36	0.23	0.18
Propylene	0.03	0.18	0.20	0.13	0.14
Butylene	0.02	0.02	0.05	0.15	0.06
Butadiene	0.01	0.01	0.03	0.04	0.04
Fuel gas	0.13	0.38	0.31	0.26	0.18
Gasoline	0.01	0.01	0.05	0.18	0.18
Gas oil	—	—	—	0.01	0.12
Pitch	—	—	—	—	0.10

carbon-hydrogen bonds and splitting of carbon-carbon bonds. There is a popular adage in ethylene plant lore, "Anything that can happen, will happen." The product from ethane cracking depends on which bond is cleaved, as shown in Figure 5–1. Ethylene forms from carbon-hydrogen fractures; methane forms from carbon-carbon cleavage with the resulting methyl radical picking up a hydrogen; even acetylene and hydrogen might form and survive. But if the olefin plant is run right, the predominant yield will be ethylene.

Propane cracking is a little more complicated because there are more combinations and permutations of possible fractures. Not only is there ethylene and methane in the outturn, but also propylene and, surprisingly, ethane. Cracking propane at a carbon-hydrogen bond can give you propylene. Cracking at the carbon-carbon bond will give you ethane and methane (after the methyl and ethyl radicals pick up hydrogens), or ethylene if both go at the same time. Ethane also gets formed as a second round draft choice when two methyl radicals find each other instead of hydrogen. Remember, "Anything that can happen . . ."

When naphtha or gas oil is cracked, imagine the limitless combinations possible. Naphthas are made up of molecules in the C_5 to C_{10} range; gas oils from C_{10} to perhaps C_{30} or C_{40}. The structure includes everything from simple paraffins (aliphatics) to complex polynuclear aromatics. So, a much wider range of possible molecules can occur. Ethylene yields from cracking naphtha or gas oil are much smaller than those from ethane or propane, as you can see from Table 5–1. But to compensate the plant operator, a full range of other hydrocarbons are produced as by-products also.

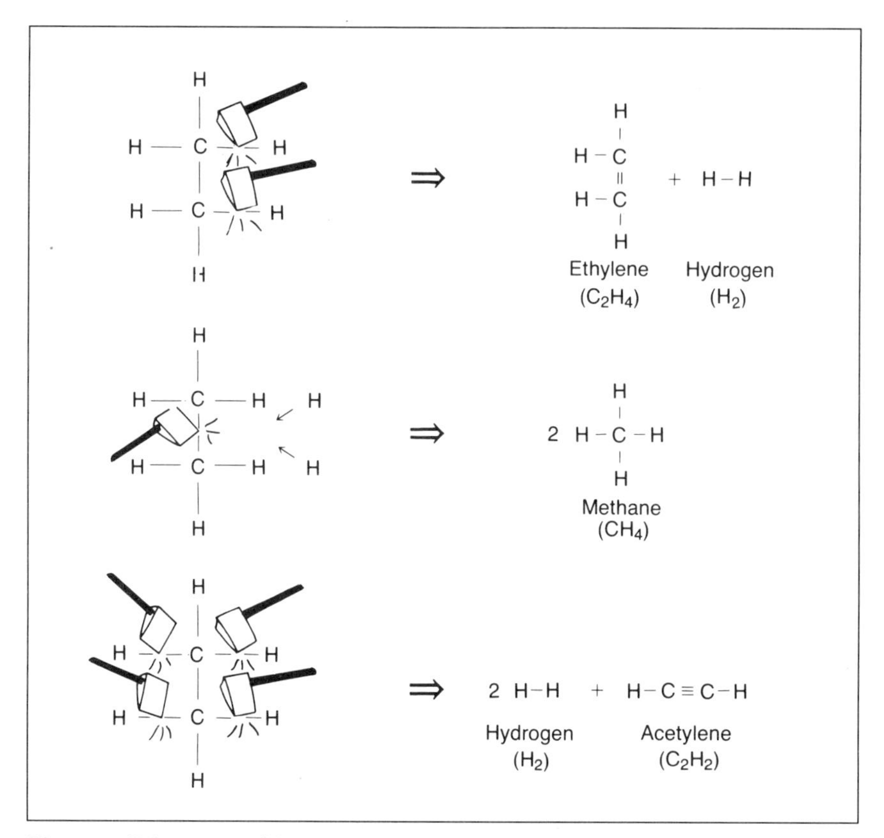

Fig. 5–1 Ethane cracking

In the C_4 and C_5 range, a new breed of molecules shows up. These are aliphatics with more than one set of double bonds. The commercially important ones are butadiene and isoprene, both of which are important raw materials for making synthetic rubber. Butadiene is big enough to be covered in much more detail in the next chapter.

The BTX's, covered in the last two chapters, result mostly from cracking away the miscellaneous chains on complicated aromatic ring containing molecules in the feed. Other hydrocarbons in the C_5/C_6 and heavier category are mostly used as refinery process or blending stocks. Many of them have higher values than the naphtha or gas oil feed. So even though they are by-products, they contribute significantly to paying for the operation of the olefin plant. In fact, in many companies, heavy liquids crackers are the link that integrates the refinery with the petrochemical plant as one complex.

THE HARDWARE

Olefin plants all have two main parts: the pyrolysis or cracking section and the purification or distillation section. The example in Figure 5–2 is an ethane cracker which has the simplest purification section. The pyrolysis (from the Greek, pyros, fire) section consists of a gas-fired furnace where the cracking takes place. The ethane is pumped through a maze of 4–6 inch diameter tubes where it is heated up to about 1500 °F and cracks. The ethane, by the way, never comes in direct contact with the fire. Otherwise, it would ignite. It stays inside the tubes.

The ethane is pumped through the pyrolysis section at a very high rate. Residence time of any individual molecules is a few seconds or less in the older plants, and less than 100 micro-seconds in some of the newer ones. This rapid rate is required to keep the cracking process from running away, resulting in the ethane cracking all the way to methane or even coke (carbon) and hydrogen. To further control this runaway cracking, the ethane is mixed with steam before it is fed to the furnaces. Steam has two beneficial effects. First, it lowers the temperature necessary for the cracking to take place; that reduces the fuel bill and also the

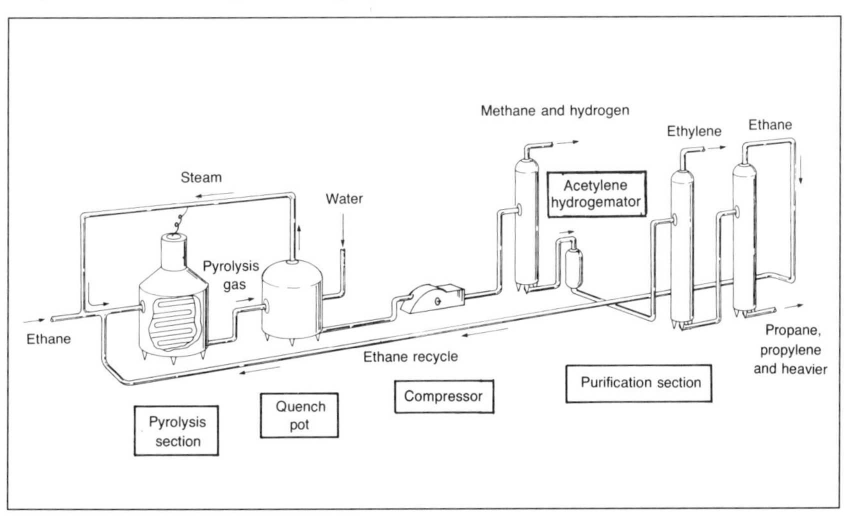

Fig. 5–2 Ethane cracker

amount of methane and hydrogen that gets formed. Second, at the lower temperature, less coke forms and deposits on the inside of the furnace tubes. That saves having to shut down the furnaces for de-coking so often, a step necessary to prevent clogging and cold spots. Coke deposits act as an insulator, preventing the ethane from heating to the right temperature for effective cracking.

As soon as the hot effluent (the gases coming out of the reaction section) leaves the cracking furnace, it enters a quench pot. The gases coming out of the furnace are so hot they will continue to crack, just like a steak will continue to cook after you take it off the grill. So the gases are immediately hit with a stream of water to cool them down. The heat transfers from the gases to the water, which causes the water to turn into steam. This steam is subsequently recycled by separating it from the effluent and mixing it with the fresh incoming feed to the furnaces.

At this point, the cracked gases consist of a mixture typically of the following composition:

	% Weight
Methane and Hydrogen	8
Ethylene	48
Ethane	40
Propane and heavier	4

You can see that only 60% of the ethane has been cracked. Forty percent of the effluent stream is still uncracked ethane. So part of the purification section will be dedicated to separating the ethane so it can be fed back to the furnaces again. This arrangement is sometimes referred to by the grizzly expression recycling to extinction. So, while the pyrolysis section only makes 48% ethylene, recycling results in the combined pyrolysis/purifications yielding:

	% Weight
Methane and Hydrogen	13
Ethylene	80
Propane and heavier	7

(and no ethane!). The transformation is just arithmetic. Just take the once-through yields, drop the 40% ethane, and divide everything by 0.60 so they add up to 100%.

In the purification section of an ethane cracker, the gas can be handled in one of two ways. In order to fractionate the streams, they must be liquified. Since they are all light gases, liquifaction can be done either by increasing the pressure in a compressor or by reducing the temperature to very low points in something called a "cold box." The ethane cracker in Figure 5–2 shows the compressor option. (Even then, the streams have to be cooled to assure they liquify.)

Downstream of the compressor is a series of fractionators (generally the tallest towers in an ethylene plant) which separate the methane and hydrogen, the ethylene, the ethane, and the propane and heavier. All are heavy metallurgy to handle the pressures and insulated to maintain the low temperatures. There's also an acetylene hydrogenator or converter in there. Trace (very small) amounts of acetylene in ethylene can really clobber some of the ethylene derivative processes, particularly polyethylene manufacture. So the stream is treated with hydrogen over a catalyst to convert little acetylene present to ethylene.

It may seem curious that an ethane cracker has propane and heavier included in the outturns. There are two reasons. The ethane used as feed is rarely pure. It generally has a couple percent of propane and heavier in it. This results in a small amount of heavier products. But why bother or go to the expense to get pure ethane feed. In the first place, the olefins plant purification section can handle them. Secondly , some heavy hydrocarbons are actually formed anyway in the frantic scramble of free radicals and hydrogen that goes on during the cracking process.

Heavier Feeds. As the feeds to the olefin plant get heavier, the hardware gets more extensive and expensive. But the flow through the plant is still about the same as for the lighter feeds, as shown in Figure 5–3. In that simplified flow diagram, the heavy liquid feed goes to the pyrolysis section where it is cracked. Next it goes to quench section where it is cooled; then to the separation section where it is split into its components.

Referring to the hardware in Figure 5–4, there are much larger facilities required for heavier liquids cracking than for ethane or propane. As you saw in Table 5–1, the yield of ethylene from the heavier feeds is much lower than from ethane. That means that to produce the same amount of ethylene on a daily basis, the gas oil furnaces have to handle nearly five times as much feed as ethane furnaces. Some of the things the

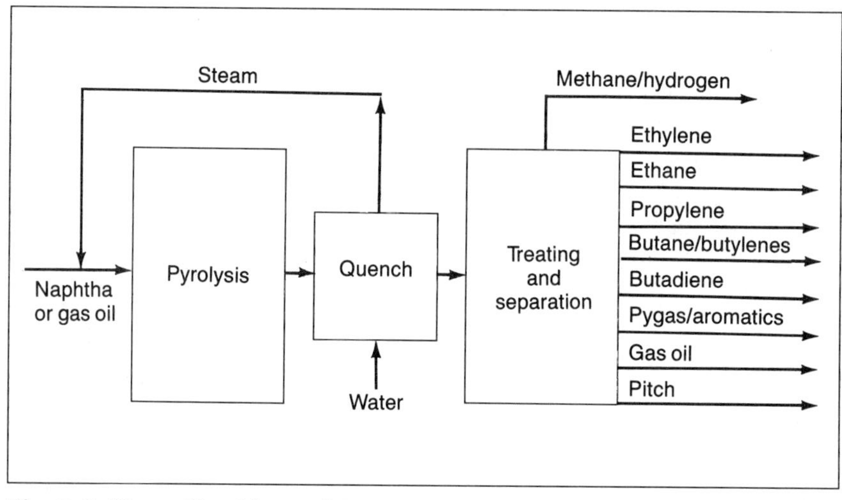

Fig. 5–3 Heavy liquids cracker

design engineer has to worry about are the size of the tubes necessary to heat up that much feed, the residence times best for each kind of feed, and the best pressure/temperature/steam mixture conditions.

The separation section of a gas oil cracker looks like a small refinery, as you can see in Figure 5–4. In addition to the fractionators and treaters used in the purification section of the simpler ethane cracker, there are facilities to separate the heavier co-products. In the front end of the separator facilities in Figure 5–4, the cold box option for handling the liquification of the gases is shown. Temperatures as low as –220°F are achieved in this super-refrigerator. At those low temperatures, freon won't do the job. Liquid air, methane, ethylene, or ammonia are often used as the refrigerant in much the same way freon is used in an air conditioner.

Several new streams are introduced in Figure 5–4. Propylene handling will be covered in detail later in this chapter. The C_4 stream is a combination of butanes, butylenes, and butadiene. Depending on the commercial interest, this mixture can be processed further to separate the individual streams. The C_5+ gasoline stream, usually called pygas (sounds like an ailment caused by pizza), is typically given a mild hydrogenation step. Some of the molecules are very reactive olefin and diolefin (two double bonds) structures which are bad actors in gasoline. They form gums and lacquers in car engines. When these olefins are

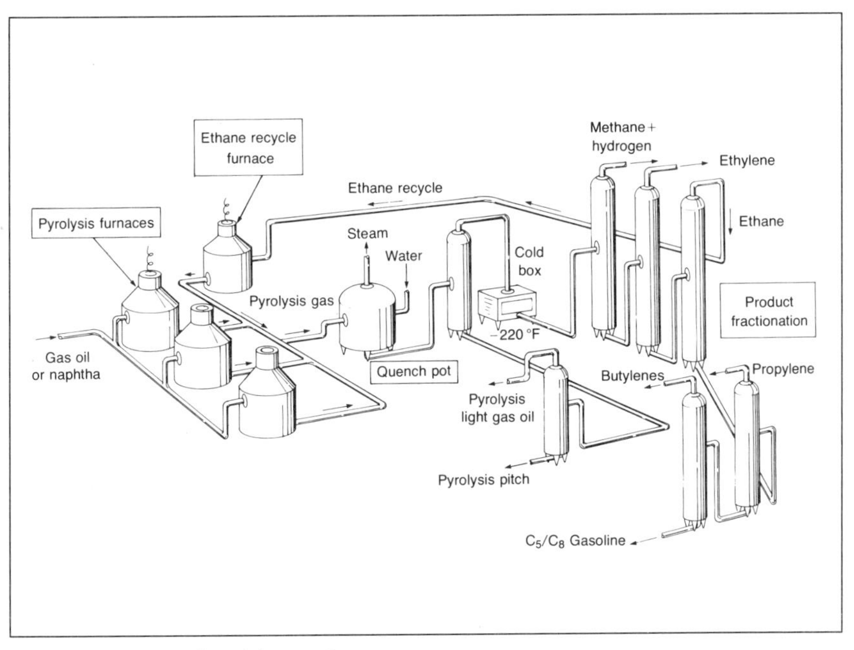

Fig. 5–4 Heavy liquids cracker

hydrogenated, the pygas stream with its high octane number becomes a good gasoline blending component. (See Chapter 2, Benzene From Olefin Plants.) It can be processed in an Aromatics Recovery Unit to remove the BTX's, which can be up to 10% of the pygas stream. But even after that it still makes good gasoline blending stock.

The fuel oils coming out of olefin plants are also characterized by an abundance of polynuclear aromatic molecules. (Same definition as for Figure 2–1.) They are sometimes inaccurately referred to as having a high aromatics content. Nomenclature aside, because of this, the burning characteristics of pyrolysis gas oil and pyrolysis pitch are poor. They are smokey, sooty, and gum formers; they tend to be more viscous, and because of their polynuclear aromatic content, they are suspected carcinogens. An unsavory set of products!

Whenever anything heavier than ethane, or sometimes heavier than propane, is cracked there is a furnace designed to handle the ethane recycle stream. The plant shown in Figure 5–4 shows three heavy liquid furnaces and one ethane furnace. Since the alternate use for ethane is usually refinery fuel, the economics often dictate recovery and cracking.

PROCESS VARIABLES

Despite the abundance of analysis in the technical journals on the subject, there really isn't much flexibility in changing the yields in olefins plants once they are built. The problem is that the yield of each of the co-products moves in a different direction as the pressures, temperatures, and residence times in the furnaces are changed. The fluctuations of the economic values of the by-products often result in little incentive to effect yield changes.

More significant, however, are the changing values of the feed-stocks. In many plants or companies, design permits substituting one feed for another, say, ethane for propane or naphtha for gas oil. In those cases, plant operations respond to the market for feedstocks and products and reflect themselves in the changing yields implied by Table 5–1.

ETHYLENE

Ethylene is a colorless gas with a slightly sweet odor. It turns from liquid to gas (boils) at –155°F. It burns readily in the presence of oxygen with a luminous flame. In fact, it was the ethylene component that made coal gas so useful as a gas light fuel at the turn of the nineteenth century. The other components in the coal gas don't give off near the light when burned by themselves. Natural gas lamps or propane/butane lanterns must be fitted with mantels to reduce the oxygen available, permitting only partial oxidation. That process gives off light. Burning ethylene doesn't require a mantel.

The logistics of ethylene are tough. Because it's a light gas, high pressures or extremely low temperatures are required to handle it as a liquid. Very little ethylene is transported by truck, and even that has to be done under special permit. For long hauls, the trucker will generally have to vent off some of the ethylene to keep the rest of it cool enough so that it can be contained at a reasonable pressure. Venting off ethylene (or any gas for that matter) is a cooling process. Ever wet your finger and stick it up in the air to see which way the wind was blowing? Whichever side of your finger got cool was where the wind was coming from. That's because the moisture on your finger was vaporizing, and that's a cooling process. In order for the moisture to go from liquid to vapor, it must pick up heat from its surroundings (your finger). Similarly, when ethylene is vented from a truck, it takes heat from the remaining ethylene. And were not the temperature kept down, the pressure of the ethylene would increase dangerously.

Ethylene trucks are expensive to build and operate. The fuel, the ethylene loss, and the energy necessary to liquify the ethylene are costly factors. So most transport of ethylene is by pipeline. Although the operating costs of a pipeline are low, the initial construction costs are high. Pipeline transport, like most other aspects of ethylene, is capital intensive. For this reason, most consumers of ethylene are located in proximity to the producers of ethylene.

Ethylene pipelines operate more like natural gas pipelines than petroleum pipelines. That is, the ethylene moves as a gas, not as a liquid. The critical temperature is the reason. There's an interesting physical phenomenon involved here. Every gas has a critical temperature, and if you keep the gas above that temperature, no matter how much you increase the pressure, the gas won't liquify. (The reason is a complex explanation having to do with atomic structure.) The critical temperature for ethylene is 48.6°F. Ethylene pipelines are usually buried about 10–15 feet below ground level, so the surrounding temperature is always 60–70°F. Ethylene, then, can be pumped around at very high pressures, 700–800 psi, and at those temperatures it is a very dense gas...nearly as dense as liquid, but still a gas.

Storage of ethylene is also an expensive proposition. For small volumes, like transfer tanks in a chemical plant, cylindrical or spherical tanks are often used. But the pressure requirements at normal temperatures demand heavy duty, thick, expensive steel vessels. Storage of any size, say beyond 100,000 pounds, warrants cryogenic storage (from the Greek kryos, cold; and gen, bring forth). Cryogenic tanks are much lighter and cheaper steel tanks. Their use is made possible because the ethylene is super-cooled way below the critical temperature (Figure 5-6). Under this condition, the ethylene is liquid and very little pressure is needed to keep the ethylene from vaporizing. The operating cost of cryogenic tanks is high. Despite the fact that there is thick insulation around the tank, some heat leaks into the ethylene. To keep the ethylene below –155°F, the boiling point, some of the vapor is drawn off the tank, passed through a refrigeration unit where it's liquified, then returned to the tank.

Fig. 5–5 Olefins plant — heavy liquids cracker. Courtesy of Shell Oil Company.

Circulating this stream faster or slower through the refrigeration unit keeps the liquid ethylene temperature in balance with the change in temperature outside the tank.

For large inventories of ethylene, in the millions of pounds, underground storage has been found very cost effective. It usually takes the form of caverns mined in rock, shale, or limestone or jugs leached out of salt in large underground salt domes as shown in Figure 5–7.

In a jug, the more common facility, ethylene is moved in and out by displacement. When ethylene is pumped in, it displaces the brine (salt water) in the jug. To remove ethylene from the jug, brine is pumped in, displacing it. Like any other hydrocarbon/water combination, the ethylene and water do not mix. So the water acts as a pressuring agent on the ethylene. Jugs or caverns are generally located a couple hundred feet below ground level. Ground temperature is a constant temperature, 65 to 70°F, and is always above the critical temperature of ethylene. So the weight of the water in the stand pipe is enough to keep the ethylene compressed and at a normal, not cryogenic, temperature for pipeline transport.

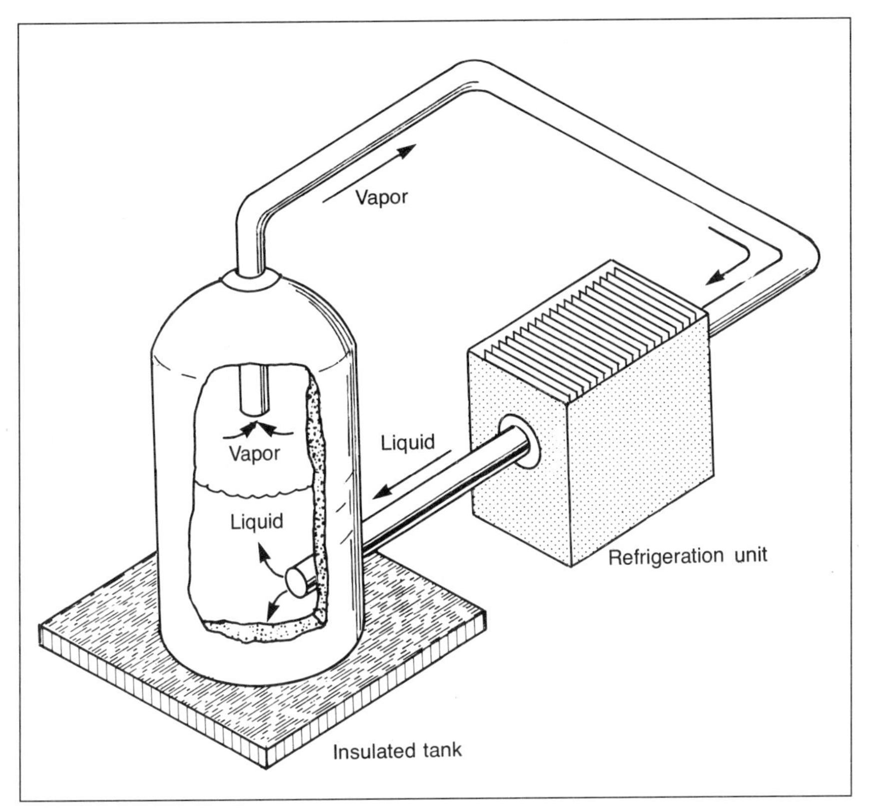

Fig. 5–6 Cryogenic ethylene storage

The cost of salt dome construction is cheaper than mined cavern storage which in turn is a lot cheaper, per pound of ethylene, than cryogenic storage and pressure storage.

The chemical uses for ethylene prior to World War II were limited, for the most part, to ethylene glycol and ethyl alcohol. After the war, the demand for styrene and polyethylene took off, stimulating ethylene production and olefin plant construction. Today's list of chemical applications for ethylene reads like the "What's What" of petrochemicals: polyethylene, ethylbenzene-styrene, ethylene dichloride, vinyl chloride, ethylene oxide, ethylene glycol, ethyl alcohol, vinyl acetate, alpha olefins, and linear alcohols are some of the more common commercial derivatives of ethylene. The consumer products derived from these chemicals are found everywhere, from soap to construction materials to plastic products to synthetic motor oils.

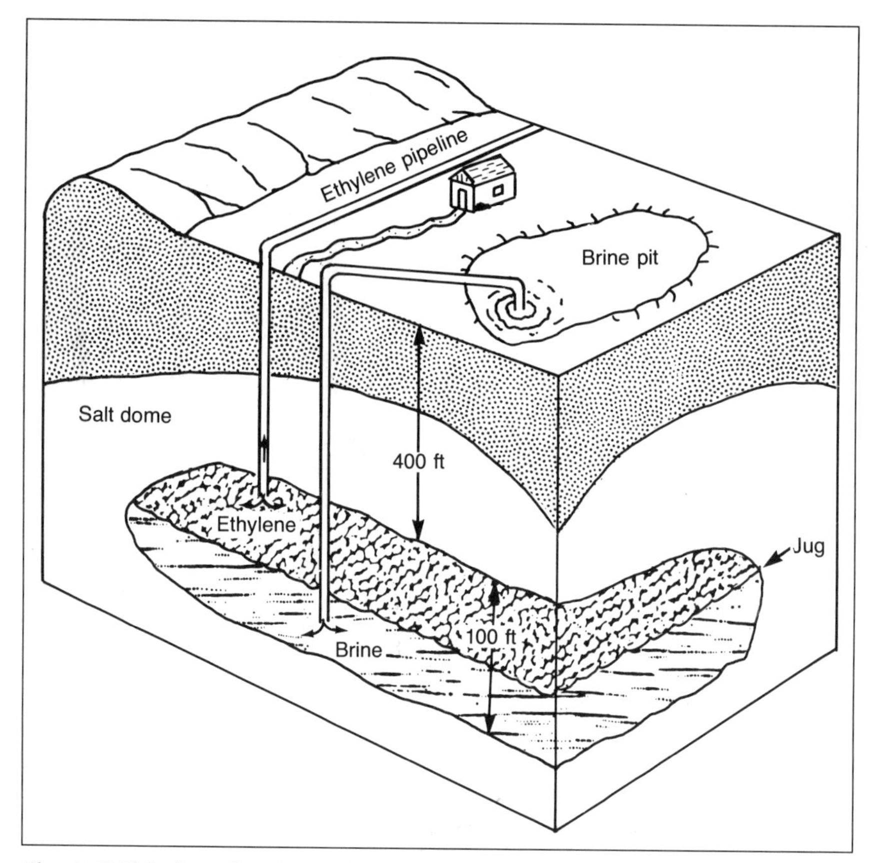

Fig. 5–7 Ethylene jug in a salt dome

PROPYLENE

Propylene, like ethylene, is a colorless gas at room temperature. It is as flammable as LPG (liquified petroleum gas or propane). In fact, propylene can be used as a substitute or supplement to LPG. The fuel characteristics are nearly indistinguishable.

Propylene is traded commercially in three grades: refinery, chemical, and polymer grade. The difference is almost entirely the ratio of propylene to propane in the stream. Refinery grade propylene usually runs about 50–70% propylene; chemical grade 90–92%; polymer grade is at least 99% propylene. The remaining percentage is almost all propane in each case.

The reasons for the three grades are very practical. For the first two, refinery and chemical, that's the way they're made. Refinery grade propylene streams are generally by-products of a refinery's cat cracker, and the propane/propylene ratio is determined by the way the cat cracker is run to make gasoline, not propylene. Chemical grade propylene is usually produced in a naphtha or gas oil cracker. The ratio of propylene and propane is 92/8 over most of the operating conditions.

Some applications, particularly polypropylene manufacture, require very pure propylene feed. Polymer grade propylene is made by simple fractionation of one of the less pure propylene streams, refinery or chemical grade.

The logistics of propylene are more conventional than ethylene, but still expensive. While ethylene is like natural gas, propylene handling and logistics are almost identical to LPG. At room temperature, propylene has to be kept in a pressurized container to keep it from evaporating. It boils at –54°F, so cooling it down to keep it liquid is expensive.

Propylene is moved in equally large volumes by pipeline, tank car, and tank truck. All three modes handle propylene as a liquid, operating at pressures of about 200 psi. The storage facilities for large volumes of propylene are the same as those for ethylene, underground jugs or caverns. Because of the lower pressure requirements than ethylene, cryogenic storage is rarely used. Storage in the form of steel spheres (typically 5–10 million pounds) and cylinders or "bullets" (200–500 thousand pounds) are prevalent.

Unlike ethylene, more propylene has always been produced than has been needed for the chemical industry. The situation goes back to the advent of thermal cracking units in refineries in the early part of the twentieth century. By World War II, with catalytic cracking units generating larger volumes of by-product propylene, chemists had been challenged sufficiently to develop both petrochemical applications and refinery uses for propylene. As a consequence, there has since been a large amount of propylene used in the manufacture of gasoline. The most popular process has been alkylation, in which a high octane C_7 hydrocarbon is made by reacting propylene with iso-butane in the presence of sulfuric or hydroflouric acid. The product is called propylene alkylate and has an octane number of about 96, so it is a good gasoline blending component.

The propylene equivalent of polyethylene is polypropylene.

About 25% of the chemical use of propylene is directed to that use. Other major applications are the manufacture of propylene oxide, isopropyl alcohol, cumene, and acrylonitrile. The consumer products you are familiar with show up everywhere: carpets, rope, clothing, plastics in automobiles, appliances, toys, rubbing alcohol, paints, and epoxy glue.

Chapter V in a nutshell...

Ethylene, C_2H_4, and propylene, C_3H_6, are both the smallest and the biggest petrochemicals. They are the largest volume petrochemicals; they have the simplest structure (at least ethylene does). Their most attractive feature is the double bond between two carbon atoms, which makes them highly chemically reactive.

Cracking large hydrocarbons usually results in olefins, molecules with double bonds. That's why the refinery cat crackers and thermal crackers are sources of ethylene and propylene. But the largest source is olefin plants where ethylene and propylene are the primary products of cracking one or more of the following: ethane, propane, butane, naphtha, or gas oil. The choice of feedstock depends both on the olefins plant design and the market price of the feeds.

In an olefins plant the feed is subjected to very high temperatures in cracking furnaces for a few moments, and then cooled rapidly to stop the cracking. Elaborate separation facilities are necessary to separate the olefins from the by-products of the cracking process.

Both ethylene and propylene are gases at room temperature and are handled in pressurized, closed systems. The list of derivatives of these two building blocks is extensive.

Exercises

1. To make at least 500 million pounds of ethylene per year and at least 200 million pounds of propylene per year, how much propane or gas oil would you have to crack in an olefins plant? How much butadiene would you make in either case?
2. Why doesn't the ethylene burn up as it passes through the furnace?
3. List all the possible products that could result from cracking butane, CH_3-CH_2-CH_2-CH_3, in an olefins plant. Remember, "Anything that can happen . . ."
4. Why do you suppose olefin plants are never referred to as propylene plants? Is propylene a second class petrochemical?

VI

THE C_4 HYDROCARBON FAMILY

"If you cannot get rid
of the family skeleton,
you may as well
make it dance."

■

**George Bernard Shaw,
1856-1950**

The first serious notice of C_4 hydrocarbons came with the development of refinery cracking processes. When catalytic cracking became popular, refiners were faced with disposing of a couple of thousand barrels per day of a stream containing butane, butylenes, and small amounts of butadiene. At first, it was all burned as a refinery fuel. The advent of the alkylation plant resulted in most of the butylenes being converted to alkylate, a high octane gasoline blending component.

During World War II, the Japanese cut off U.S. access to sources of natural rubber, giving the Americans a strategic imperative to develop and expand the manufacture of synthetic rubber. The C_4 streams in refineries were a direct source of butadiene, the primary synthetic rubber feedstock. Almost as a matter of serendipity, the availability of this stream was growing rapidly with the expansion of catalytic cracking to meet wartime gasoline needs. Additional butadiene was manufactured by dehydrogenation of butane and butylene also.

In the 1950s, the olefin plants that were cracking propane and butane, began to produce modest amounts of by-product C_4 hydrocarbon streams. Later in the 1960s and 1970s, gas oil- and naphtha-based olefins plants rivaled refineries in the volume of the C_4 streams being produced.

A typical C_4 hydrocarbon stream coming from a gas oil or naphtha cracker like that shown in the last chapter in Figure 5–4 might have the following composition:

Iso-butane	5%
Normal butane	5%
Butadiene	42%
Iso-butylene	18%
Butene-1	18%
Butene-2	12%

The terms butene-1 and butene-2 are the petrochemical term used to refer to what the petroleum refining industry calls normal butylenes. Butene-1 and -2 are more specific and descriptive. But to complicate matters more, there are two kinds of butene-2: cis-butene-2 and trans-butene-2. In Figure 6–1, if you look closely, you'll see that the difference between butene-1 and the butene-2's is the location of the double bond. Butene-1 has it at the end position; butene-2 at the middle. The methyl groups in trans are across from each other, on opposite sides of the fence; in cis they are next to each other or on the same side of the fence. The difference is more than cosmetic. It determines the way the molecule behaves, physically and chemically. Check the boiling points, for instance, in Figure 6–1. They're different, and that helps in the separation process. In a few paragraphs the different applications that derive from the chemical behavior differences will be discussed.

The structural difference between the two butadienes is pretty obvious in Figure 6–1. But most mixtures of butadiene are predominantly

1,3 butadiene, and there is little attention paid to the difference between the two.

Configuration	Name	Boiling Temperature, °F
$CH_3-CH(CH_3)-CH_3$	Isobutane	10.9
$CH_3-C(CH_3)=CH_2$	Isobutylene	19.6
$CH_3-CH_2-CH=CH_2$	Butene-1	20.7
$CH_3-CH=C=CH_2$	1, 2 butadiene	51.4
$CH_2=CH-CH=CH_2$	1, 3 butadiene	24.1
$CH_3-CH_2-CH_2-CH_3$	Normal butane	31.1
$(CH_3)(H)C=C(H)(CH_3)$ (trans)	Trans-butene-2	33.6
$(CH_3)(H)C=C(CH_3)(H)$ (cis)	Cis-butene-2	38.7

Fig. 6–1 Characteristics of the C_4 Hydrocarbons

PROCESSING

There are a dozen different ways to handle the C_4 stream in a petrochemical plant if you follow all the combinations possible in Figure 6–2. Simple fractionation won't do it because the boiling temperatures are so close together. Generally the first step is to remove the butadienes by extractive distillation, of the kind shown in Chapter 3.

Iso-butylene is the most chemically reactive of the butylene isomers. If the objective is just to get the isobutylene out of the C_4 stream, it can be removed by reaction with methanol (CH_3OH) to make MTBE (methyl tertiary butyl ether), by reaction with water to make TBA (tertiary butyl alcohol), by polymerization, or by solvent extraction. After that, butene-1 can be removed by selective adsorption or by distillation. That leaves the butene-2 components, together with iso- and normal butane, which are generally used as feed to an alkylation plant.

Not all chemical plants have all these facilities. Furthermore, some plants have processes to convert butenes to butadienes; others convert them the other way. The best way to sort out the options is to treat them one at a time.

BUTADIENE

Butadiene is used primarily as a feedstock for synthetic rubber, elastomers, and fibers. It has grown to be a major petrochemical building block and commodity. Butadiene is a colorless gas at room temperature, but it is normally handled under pressure or refrigerated as a liquid.

The base-load supply of butadiene is from olefins plants simply because butadiene is co-produced with the other olefins. There's not much decision on whether or not to produce it. It just comes out, but in a small ratio compared to ethylene and propylene. Cracking ethane yields one pound of butadiene for every 45 pounds of ethylene; cracking the heavy liquids, naphtha or gas oil, produces one pound of butadiene for every seven pounds of ethylene. Because of the increase in heavy liquids cracking, about 75% of the butadiene produced in the U.S. is co-produced in olefin plants.

The swing supply, or "on-purpose," butadiene is made by catalytically dehydrogenating (removing hydrogen from) butane or butylene.

Processes

The processes specific to butadiene are the butane/butylene dehydrogenation and the butadiene recovery operation.

Dehydrogenation. The process for making butadiene from butane or butylene involves passing the feed over a catalyst at about 1200°F and reduced pressure. Depending on the feed, two or four hydrogen atoms pop off the butane (C_4H_{10}) or butylene (C_4H_8) molecules, forming butadiene (C_4H_6).

The catalysts used are metal oxides or phosphates—ferric oxide, chromic oxide, or calcium-nickel phosphate. After one pass, the effluent usually goes through the extractive distillation process.

Recovery. The extractive distillation process for removing butadienes from the C_4 stream uses a solvent that reduces the boiling point of the butadiene. The C_4 stream is fed to the middle of a fractionator, and a high boiling point solvent is fed at the top. The solvent, as it works its way down, strips out the butadiene as the C_4 vapor works its way up the column. The solvent and butadiene come out the bottom and can easily be split in a second column. Two popular high boiling point solvents are N-methylpyrrolidone (NMP) and dimethylformamide (DMF).

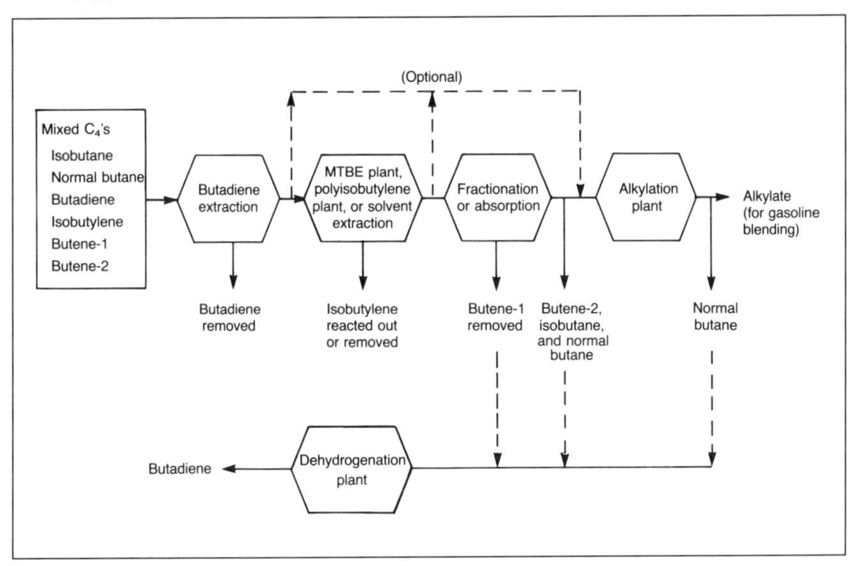

Fig. 6–2 Typical C_4 processing scheme

End Uses Markets

Butadiene's two double bonds make it very reactive. It readily forms polymers, reacting with itself to form polybutadiene. Also, it's used as a co-monomer to make styrene-butadiene rubber (SBR), polychloroprene and nitrile rubber. These are all forms of synthetic rubber and account for about 70% of the butadiene consumed. The largest share of them is on highway vehicles—truck and car tires, hoses, gaskets, and seals.

Some of the non-rubber applications are adiponitrile and hexamethylenediamine, precursers to making Nylon 66. Carpeting is the primary application of Nylon 66. Other non-rubber applications are styrene-butadiene latexes for paper coatings and carpet backing, and acrylonitrile-butadiene-styrene (ABS) resins for plastic pipe and automotive/appliance parts.

ISOBUTYLENE

The isobutylene in the C_4 stream generally ends up in one of four places: an MTBE plant, a polymerization process, a solvent extraction process, or in a refinery alkylation plant. The first three are methods of removing isobutylene from the C_4 stream; the fourth is the "default." The isobutylene just follows the normal butenes to a process for making gasoline.

MTBE. The major use of isobutylene as a separated petrochemical is to make MTBE, a gasoline blending component with two meritorious attributes. It has a high octane, and it has oxygen in its molecular structure. While the former was enough to get interest in MTBE started, the latter has increasing appeal due to its environmental implications. The presence of the oxygen in the molecule facilitates ozone-free combustion of gasoline in a vehicle, eliminating a potential air pollutant.

The MTBE process also is attractive because the entire C_4 stream can be fed to it. The isobutylene is selectively reacted out.

Polymerization. Similarly, the polymerization process will pull the isobutylene selectively out of the C_4 stream. Polyisobutylenes are used mainly as viscosity index improvers in lubricating oils and as caulking and sealing compounds. Some of the low molecular weight polyisobutylenes are particularly suited for use in the construction field because they don't solidify. They remain a tacky fluid and when properly formulated with clay fillers, etc., take on the properties of a sticky, putty-like substance.

Solvent Extraction. The segregated isobutylene stream from the solvent extraction process is used or sold for use in numerous applications, besides the polyisobutylenes just mentioned: butyl rubber, alkylated aromatics, oxo alcohols, tertiary butyl alcohols, di- and tri-isobutylenes and methyl methacrylate.

Processes

In an MTBE plant, isobutylene is selectively reacted with methanol over a catalyst to produce methyl tertiary butyl alcohol, as shown in Figure 6–3.

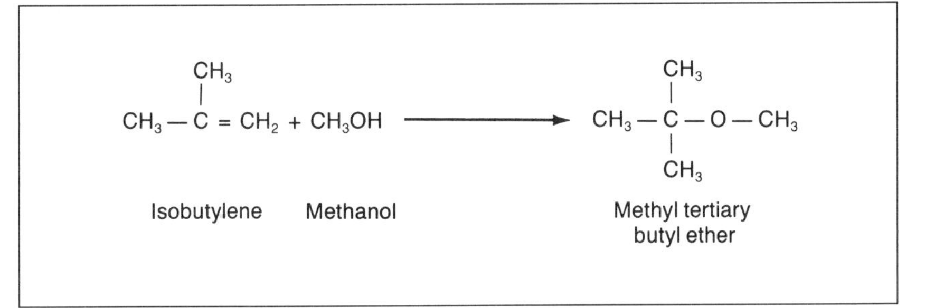

Fig. 6–3 MTBE reaction

The reaction takes place in a fixed bed reactor. The process is exothermic, so cooling coils must remove the heat from the reactor continuously. The reactor effluent contains unreacted feeds and MTBE and is fractionated into recycle streams and product.

The polymerization process is a low temperature catalytic reaction. The type of polymer produced is strongly affected by the reaction temperature. Low temperatures give low molecular weight polymers, the kind useful in caulking compounds and as viscosity index improvers for motor oils.

A high purity isobutylene stream can be recovered by solvent extraction using cold sulfuric acid. One problem occurs if there is any butadiene left in the stream—sulfuric acid will cause it to polymerize. But if the butadiene has been first extracted, a 99+ percent isobutylene stream can be recovered.

BUTENE-1

The demand for high purity butene-1 (called polymer grade) rapidly developed in the 1970s. Butene-1 was always a popular comonomer with ethylene in high density polyethylene (HDPE). But the rapid growth of linear low density polyethylene (LLDPE) starting in the 1970s increased the demand for butene-1 from 10 million to 500 million pounds per year in the next decade and a half.

Other minor petrochemical uses of butene-1 continue to be the manufacture of SBA (secondary butyl alcohol), MA (maleic anhydride), polybutylene for piping applications, and butylene oxide.

The boiling points of butadiene, iso-butylene and butene-1 make it impractical to recover a high purity butene-1 stream without first removing the other two by methods other than fractionation, as covered above. After that, the butene-1 still needs to be separated from the other C_4's, but that can be by fractionation. That's still an expensive proposition because the boiling temperatures of iso-butane, normal butane, and butene-2 are not all that different. An alternate route, molecular sieve adsorption, works well.

The Processes

The applications of butene-1 usually require very low levels of isobutylene and butadiene. Sometimes an extra reactor in the MTBE plant is added to get the isobutylene content down from the typical 2.0% to a 0.2% level. Small amounts of butadiene are removed by hydrotreating the stream over a catalyst, which converts the butadiene to butene-2, and maybe some butane.

Distillation. The distillation method of separating butene-1 is difficult. It requires a column with over 100 trays operating under a reflux ratio of about 150:1. (Chemical engineer jargon—it means a very tall column, with lots of recycle—very energy intensive.)

Adsorption. The second technology is selective adsorption. The use of molecular sieves was discussed in the chapter on xylenes. Molecular sieves, you will recall (of course) are crystals with millions of pores, all of a uniform size or shape. In this process, a sieve with pores that will fit only butene-1 is used.

The process runs on a cycle. First the C_4 stream is fed to a vessel packed with the molecular sieve. The butene-1 molecules start to fill up the sieve's pores. After a while, when the pores are about saturated, the feed is cut off. Another liquid, the desorbent, is flushed back through the vessel, and the butene-1 is washed out of the sieves. The desorbent is selected so that after it picks up the butene-1 from the sieve, it can easily be separated from the butene-1 by fractionation. The key, of course, is to use a desorbent with a boiling temperature a good distance away from butene-1's. Any run-of-the-mill hydrocarbons that fit this criteria are suitable.

BUTENE-2

At the end of the line, after all the goodies that anyone wants have been removed, is the alkylation plant. That's where most of the butene-2 goes, because there are few chemical applications for this molecule. The only thing good you can say about butene-2 is that it makes a better (higher octane) alkylate than butene-1.

Alkylation. The term alkylation generally applies to the addition of an olefin to a branch chain hydrocarbon. In a refinery alkylation plant, isobutane is the branch chain; the olefins are propylene and the butylenes. The result is a branched C_7 or C_8 hydrocarbon with good motor gasoline characteristics called alkylate. The alkylation of isobutane with isobutylene produces iso-octane which has an octane rating of 100. That's not just a random occurrence. Iso-octane is the compound that was originally used in its pure form to set the definition of octane rating.

The Process

Mixed C_4 or C_3 streams are fed to a reactor, together with an excess of isobutane (about a 12 to 1 ratio). The reactors contain cold sulfuric or hydroflouric acid which acts as a catalyst. Active mixing along with a long residence time of 15–20 minutes results in reaction of the olefins with the isobutane. The propane and normal butane that is typically present with mixed C_3 and C_4 streams are unaffected by the process and just float on through. Distillation at the tail end of the plant easily separates the C_7 or C_8 alkylate from the propane, normal butane, and the isobutane, which is recycled to the reactor. Quality-wise, the alkylate made from butenes is better (higher octane) than that from propylene.

Chapter VI in a nutshell...

The C_4 family includes normal and isobutane, C_4H_{10}, which have only single bonds; normal and isobutylene, C_4H_8, which each have one double bond; and butadiene, C_4H_6, which has two double bonds. The reactivities and chemical versatility of these three groups are roughly related to the number of double bonds.

The source of these compounds is varied. The butanes are found naturally in crude oils and natural gas. They, plus the olefins are products of various refinery processes and of olefins plants. They are separated by fractionation, except for butadiene, which is usually recovered by solvent extraction. They all vaporize at room temperature, so they are handled in closed, pressurized systems.

The butanes are used as gasoline blending components. Normal butane is sometimes an olefins plant feed. Isobutane is used in refinery alkylation plants with propylene or butylene to make alkylate, a high octane gasoline blending component.

Butene-1, normal butylene with the double bond between the end and the second carbon, is used as a comonomer in making polyethylene. Polybutylene and polyisobutylene are the polymers. Butadiene is used to make complex polymers, including synthetic rubbers.

Exercises

1. Why do some people convert butylenes to butadiene, but others convert butadiene to butylene?
2. Which do you suppose is the swing supply of butadiene:
   a. olefins plant
   b. refinery
   c. butane dehydrogenation
   d. butylene dehydrogenation
   e. Swinging Sal's butadiene shop
3. Describe how you'd use the following processes to handle the C_4 stream in the heavy liquids cracker in Figure 5–3:
   a. extractive distillation
   b. fractionation
   c. adsorption
   d. polymerization
   e. dehydrogenation
   f. alkylation

COMMENTARY ONE

"This is not the end. It is not even the beginning of the end. But it is, perhaps, the end of the beginning."

■

Winston Churchill (1874–1965), commenting on November 10, 1942 about the victory at Alamein

REVIEW

After five chapters of product and process descriptions, you might be dismayed to find out it all can be summarized in one diagram (see following figure). At least the essential parts are there, and they can provide you a quick reference.

The petrochemical products from olefins plants are ethylene, propylene, C_4's (butanes, butadiene, and the butylenes), and a stream containing BTX's. Refinery cat crackers produce propylene and C_4's. They also produce some ethylene, but often it is not recovered.

The propylene from olefins plants is usually chemical grade—from cat crackers, refinery grade. Both can be upgraded to polymer grade by fractionation.

Refinery cat reformers produce a reformate

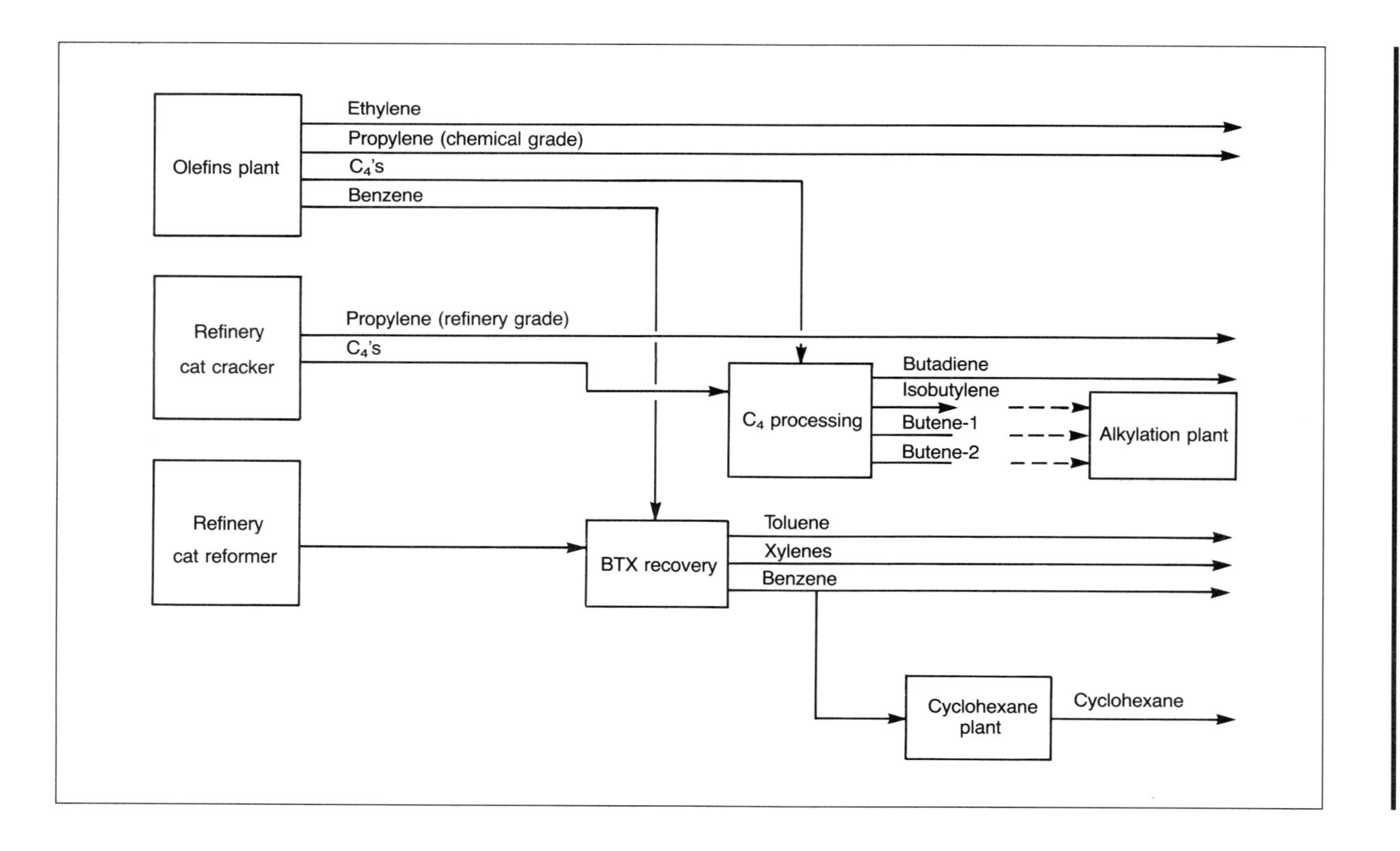
Olefins plant
Ethylene
Propylene (chemical grade)
C_4's
Benzene
Refinery
cat cracker
Propylene (refinery grade)
C_4's
Refinery
cat reformer
C_4 processing
Butadiene
Isobutylene
Butene-1
Butene-2
Alkylation plant
BTX recovery
Toluene
Xylenes
Benzene
Cyclohexane
plant
Cyclohexane

stream laden with aromatics. That stream, with or without the benzene-laden stream from the olefins plant, can be split apart in the various processing schemes in the BTX recovery facility.

The C_4's can be recovered in numerous ways to make petrochemical feeds. Often everything but the butadiene ends up in an alkylation plant.

Benzene can be hydrogenated to cyclohexane directly.

FOREWORD

The next five chapters cover the "Mutt and Jeff" petrochemicals. They're grouped together because they come in pairs. You don't have one without the other. The metaphor is accurate for cumene and phenol, ethyl benzene and styrene, and ethylene dichloride and vinyl chloride. It's a little strained with ethylene oxide and ethylene glycol and with propylene oxide and propylene glycol because there are things you can do with the oxides other than make the glycols. But there's virtually no other use for cumene, ethyl benzene, and ethylene dichloride than to use each to make its buddy.

Anyway, these 10 chemicals are first and second generation derivatives of the basic building blocks and are important commodity petrochemicals.

VII

CUMENE AND PHENOL

"Every man serves a useful purpose: a miser, for example, makes a wonderful ancestor."

■

Laurence J. Peter

The only reason petrochemical companies make cumene is to use it to make phenol. There are other ways to make phenol, but not much other commercial use for cumene.

CUMENE

During World War II, isopropyl benzene, more commonly and commercially known as cumene, was manufactured in large volumes for use in aviation gasoline. The combination of a benzene ring and an iso-paraffin structure made for a very high octane number at a relatively cheap cost. After the war, the primary interest in cumene was for the manufacture of cumene hydroperoxide. This compound was used in small amounts as a catalyst in an early process of polymerizing butadiene with styrene to make synthetic rubber. It was only by accident that it was discovered that mild treating of cumene hydroperoxide with acid resulted in the formation of phenol and acetone. Serendipity is not uncommon in the discovery process involving petrochemicals.

The Process

The reaction of benzene with propylene produces cumene (Figure 7–1), but a catalyst must be present to make the reaction go. The chemistry is such that the benzene-propylene bond will be at the middle carbon of the propylene molecule, hence, the name isopropyl benzene. Note that there is also a transferal of hydrogen from benzene to the propylene.

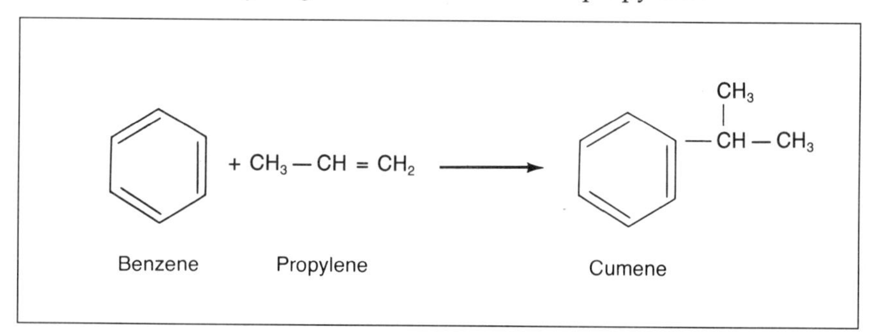

Fig. 7–1 Benzene-propylene route to cumene

The reaction can be carried out with the benzene and propylene in either the liquid or vapor phase; the more common process is vapor phase, carried out at about 425 °F and 400 psi.

The process diagram in Figure 7–2 shows propylene and benzene being fed directly to the reactor. Usually, chemical grade propylene is used because the presence of the 6–10% propane does not affect the reaction. A depropanizer can be used if a refinery grade propylene stream is used.

The reactor is a vessel with beds of solid catalyst. Most commercial processes use a catalyst called kieselguhr, which is phosphoric acid deposited on a silica/alumina pellet. Because of the weight of the pellets, supported beds at multiple levels in the vessel are used so the bottom layers won't be crushed.

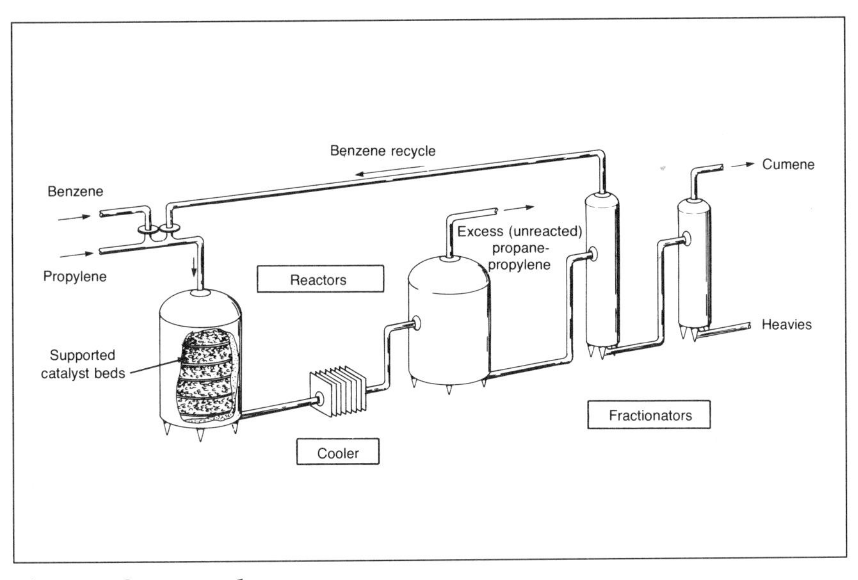

Fig. 7–2 Cumene plant

Two vessels are used for the reaction, for two reasons. First, the reaction is exothermic and, in a fixed catalyst bed, one way to control the temperature is to take out the streams being processed and cool them down. The second reason is that the second reactor also is used as a fractionator, venting the unreacted propylene and the propane part of the chemical grade propylene from the benzene/cumene mix.

Excess benzene is always used in the reactors, also for two reasons. First, the benzene acts like a heat sponge, mitigating the rate at which the temperature increases due to the exothermic reaction. Second, excess benzene helps eliminate some of the undesirable side reactions that can take place, mainly the formation of di- or tri-isopropyl benzene (benzene hooking up with two or three propylenes) or other miscellaneous compounds.

So the streams coming out of the reactors will be a mixture of the excess benzene and the product, cumene. A fractionating column is used to separate the two, permitting the benzene to be recycled. A final fractionator takes the cumene overhead; any of those miscellaneous compounds accidentally formed in the process go out the bottom.

Cumene made in this manner is about 99.9% pure. The cumene yield, i.e., the percent of benzene that ends up as cumene, is about 95%.

About 5% of the benzene ends up as part of the heavies. Conversion of propylene is a little lower, about 90%, particularly if there's no depropanizer up front to which the unreacted propane/propylene from the second reactor can be recycled.

Material Balance

Feed:	
Benzene	681 lbs.
Propylene	367 lbs.
Catalyst	small amount
Product:	
Cumene	1000 lbs.
Heavies	48 lbs.

Occasionally you might come across a compound called pseudo-cumene, which is a benzene ring connected to three methyl groups. This compound is an isomer of cumene known as 1,2,4-trimethyl benzene. Pseudo-cumene is a starting material for the manufacture of trimilletic anhydride, an important ingredient in alkyl resin paints and high temperature aerospace polyimide resins.

Commercial Aspects

Cumene is a colorless liquid, soluble in benzene and toluene and insoluble in water. It can be shipped in tank cars, tank trucks, barges, and drums. The flash point is high enough that it is not considered a hazardous material, and no DOT red shipping label is required.

Cumene Properties

Freezing point	−140.8°F (−96°C)
Boiling point	306.5°F (152.5°C)
Specific gravity	0.8632 (lighter than water)
Weight per gallon	7.19 lbs./gallon

The grades used in commerce are Technical (99.5% concentration), Research (99.9%) and Pure (only trace impurities). About 97% of production is presently Technical grade.

PHENOL

Phenol has been used for decades in the medical field as an antiseptic under its alias, carbolic acid, and at one time as a preservative of human organs under the name creosote (from the Greek kreos, flesh; and sogein, to preserve). The name creosote eventually became associated with the wood preservative, but phenol remains the principal ingredient in this product.

The early sources of phenol were the destructive distillation of coal and the manufacture of methyl alcohol from wood. In both cases, phenol was a by-product. Recovered volumes were limited by whatever was made accidentally in the process. Initial commercial routes to "on-purpose" phenol involved the reaction of benzene with sulfuric acid (1920), chlorine (1928), or hydrochloric acid (1939); all these were followed by a subsequent hydrolysis step (reaction with water to get the -OH group) to get phenol. These processes required high temperature and pressures to make the reactions go. They're multi-step processes requiring special metallurgy to handle the corrosive mixtures involved. None of these processes is in commercial use today.

In 1952, a technological breakthrough was found, the cumene oxidation route. It was much cheaper and it quickly proliferated, and is now the primary route, accounting for 97% of today's phenol production. Despite the fact that the additional step of making cumene was involved, the less severe operating conditions throughout (pressures, temperatures, acid strength) were sufficient offsets to make the process economically attractive.

Cumene Oxidation Process

This two-step process involves oxidation of cumene to cumene hydroperoxide, which decomposes with the help of a little dilute acid into phenol and acetone, as shown in Figure 7–3.

In the first step, cumene is fed to an oxidation vessel, as shown in Figure 7–4, where it is mixed with a dilute aqueous sodium carbonate solution (soda ash with a lot of water). A small amount of sodium stearate is added and the whole mixture becomes an emulsion.*

The purpose of the cumene emulsion is to permit good contact of the cumene with oxygen. The oxygen is introduced as air in the bottom of the vessel and bubbled through the emulsion. As it does, the cumene converts to cumene hydroperoxide, as shown in step 1 in Figure 7–3.

*Emulsions abound in everyday use. Mixing flour and water with meat juices make the emulsion called gravy. Putting soap powder in a washing machine causes the dirt in clothes not only to be removed but become suspended in the water—an emulsion. The dirt particles remain in this state until they're rinsed away by draining the washing machine. Seventy-five years ago the emulsion was lye. Before that, washing was mechanical—rubbing on a washboard, or even a rock.

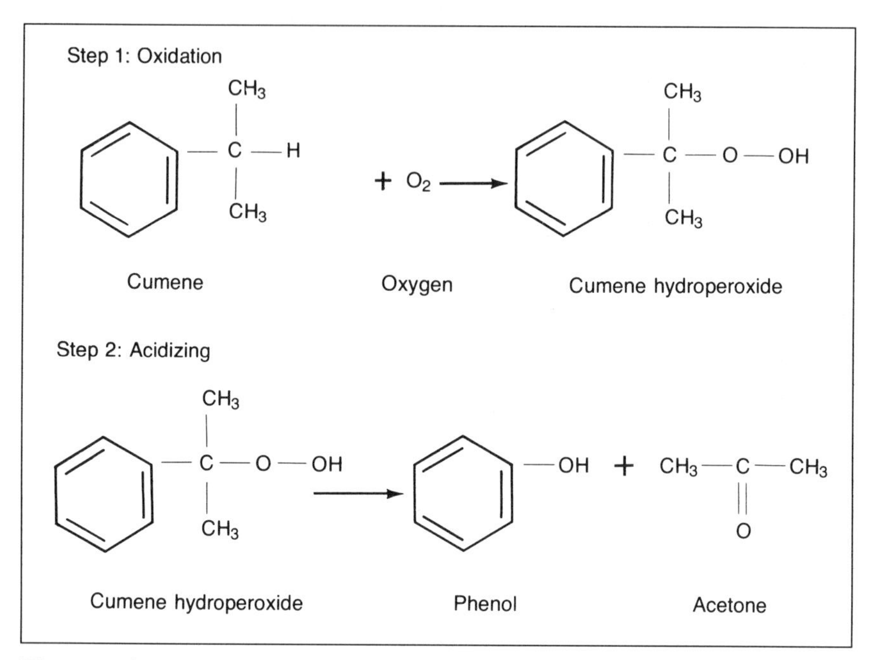

Fig. 7–3 Cumene-to-phenol process

This chemical reaction, like most oxidations, is exothermic, so it generates heat and is susceptible to a runaway reaction, sometimes called the "begets." Rapidly increasing temperatures begets an increased rate of reaction which begets more rapidly increasing temperatures which begets . . . The presence of the excess water sponges up some of the heat and reduces the risk of the begets.

To further control the runaway risk, the reaction temperature is kept at about 250°F by regulating the reactor flow-through rate. At that temperature only about 25% of the cumene is converted to phenol. So the stream coming out the bottom of the oxidizing vessel is 25% cumene hydroperoxide and 75% unconverted cumene.

At the top of the vessel (Figure 7–4) is the necessary plumbing to take off the nitrogen content of the air, which just passes through untouched, plus any excess oxygen not used in the reaction.

The bottom stream is fed to a fractionator to split out the unreacted cumene in order to recycle it to the oxidizing vessel. The cumene hydroperoxide, now concentrated to about 80%, is fed to another vessel for the second reaction step. The chemical trick here is to chop out

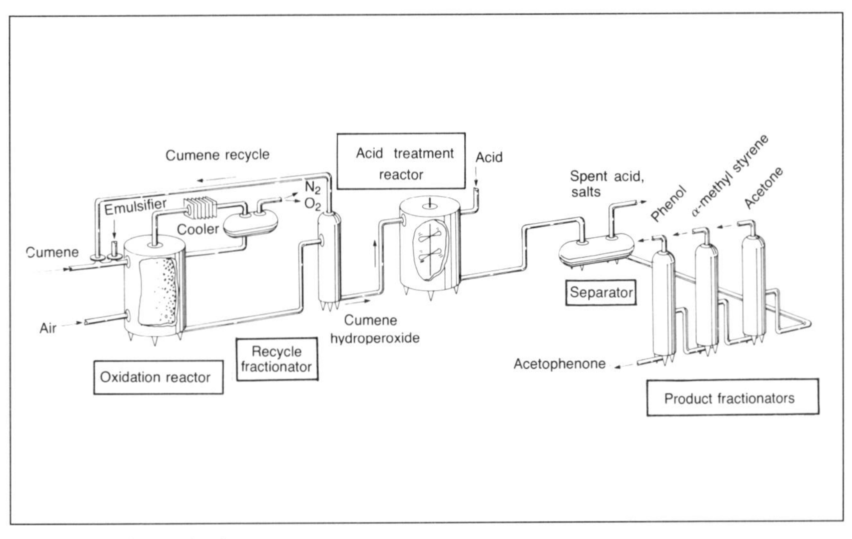

Fig. 7–4 Phenol plant

everything in between the benzene ring and the -OH group, as shown in step 2 of Figure 7–3. The use of dilute sulfuric acid does the job by initiating an unusual decomposition involving actual migration of the benzene ring around the cumene hydroperoxide molecule. The decomposition and the resulting products were a complete surprise when they were discovered. Like so many other petrochemical processes, serendipity played a role in generating progress in this chemical.

To facilitate the reaction, the mixture is stirred vigorously. It takes place at about 160–175 °F and 25–50 psi, so the conditions here, like in the oxidizer, are not too severe or expensive.

The effluent from the acid treatment reactor is about 60% phenol, 35% acetone, plus some miscellaneous cats and dogs, most of which are alpha-methyl styrene and acetophenone. The effluent is passed through a separator where the acid, water, and salts drop out. The balance of the processing is a series of distillation columns which split out the various streams.

The alpha-methyl styrene is usually catalytically treated with hydrogen and converted back to cumene for recycling. The acetophenone has some commercial use in pharmaceuticals and at one time was used to make ethylbenzene. A high purity phenol is sometimes made by a crystallization step, since phenol freezes at about 109 °F.

Material Balance	
Feed:	
Cumene	1000 lbs.
Oxygen (in excess)	300 lbs.
Sodium carbonate	small
Sulfuric acid	small
Product:	
Phenol	725 lbs.
Acetone	442 lbs.
By-products	99 lbs.
Unused oxygen	34 lbs.

Other Routes

There are a few other commercial phenol processes that deserve some mention. One involves oxidation of toluene using a cobalt catalyst to give benzoic acid, followed by a reduction (removal of an oxygen atom) to give phenol and carbon dioxide. But the process has had its headaches and has not proliferated.

Another route is the direct oxidation of benzene, a chemist's dream, but this method is difficult to control—lots of side reactions, etc.—and has not been successful.

The third route is a little more popular. Cyclohexane is oxidized to cyclohexanol, which is then dehydrogenated to phenol.

In addition to "on-purpose" phenol, very small amounts of "natural" phenol are recovered in petroleum refining operations . . . about 1% of the total supply. Almost all of it comes from caustic washing of cat cracker streams.

The cumene process remains the economic route to phenol. The only drawback is the co-production of acetone, and that only becomes a problem when acetone is long and phenol is short. What's to be done with the excess acetone production? It tends to be put into the market, undercutting the "on-purpose" acetone production. The reverse situation occurs when acetone is short and phenol is long. When both phenol and acetone are short, running hard to make phenol is not a problem.

Commercial Aspects

Uses. The major applications of phenol are phenolic resins, Bisphenol A, and caprolactam. The reaction of phenol with formaldehyde gives liquid phenolic resins (used extensively as the adhesive in plywood) and solid resins (used as engineering plastics in electrical applications). In powder form the phenolic resin can be molded easily and are completely non-conductive. These phenolic resins or plastics can be found in panel boards, switch gears, and telephone assemblies. The agitator in your washing machine is probably a phenolic resin.

Phenol also is used to manufacture several important monomers. Bisphenol A, a phenol derivative, is used to make very strong polycarbonate plastics and epoxy resins (the kind you buy in two tubes and mix to make glue). Other applications of epoxy resins include paints, fiberglass binder, and construction adhesives.

About half the caprolactam is made from phenol. (The other half comes from cyclohexane.) Caprolactam is an intermediate step to making Nylon 6.

Other miscellaneous derivatives of phenol include non-ionic detergents, aspirin, disinfectants (pentachlor phenol), adipic acid (a Nylon 66 intermediate), and plasticizers.

Properties. Phenol is a solid at room temperature and is usually handled as a powder. In its pure form, it is white in color, but exposure to sunlight or air will cause it to turn reddish. Phenol is and acts like acid. It burns, it's corrosive, and it has an odor and taste that will knock you over—literally. It's a Class B poison.

Phenol Properties

Freezing point	109.4°F (43°C)
Boiling point	359.2°F (181.8°C)
Specific gravity	1.071 (heavier than water)
Weight per gallon	9.0 lbs/gal

Phenol can be shipped in liquid form in lined tank cars or tank trucks or in galvanized drums. It's imperative that it be handled in closed systems as it will absorb water from the atmosphere. In powder form, phenol will absorb enough water from the atmosphere to turn itself into a liquid.

The powder form of phenol is usually traded either as a U.S.P. (98% minimum) or as a C.P. or synthetic grade (95% minimum).* In the liquid form, the commercial grades are 90-92 percent purity and 82-84 purity).

*U.S.P. (United States Pure) and C.P. (Chemical Pure) are nomenclature from the pharmaceuticals industry, the former indicating a grade suitable for human consumption or for manufacture of a consumable.

Chapter VII in a nutshell...

Cumene, $C_6H_5CH(CH_3)_2$, is a benzene ring with a unique functional group hung in it in the place of a hydrogen atom. It is made by reacting benzene with propylene. Fifty years ago it was used primarily as a high octane aviation gasoline component, but it is now used almost entirely as feed to coproduct manufacture of phenol and acetone.

Phenol, C_6H_5OH, is a benzene ring with a hydroxyl group, -OH, in place of a hydrogen. That makes it a member of the alcohol family. At room temperature phenol is a solid but is corrosive like an acid. It is used to make phenolic resins and to make Bisphenol A (feed for epoxy and polycarbonate resins) and caprolactam (feed for Nylon 6).

Exercises

1. The Cutthroat Phenol Company is thinking about starting up and selling out their 50 million pound per year phenol plant. The market for the various feedstocks and products is as follows. (F.O.B. the Cutthroat plant):

Benzene	\$ 1.60/gal.
Propylene	0.25/lb.
Oxygen	0.50/lb.
Acetone	0.60/lb.
By-products	0.10/lb.

   Their cumene plant operating costs are \$0.20/lb. of feed; the phenol plant operating costs are \$0.25/lb. of phenol.

   How much should Cutthroat charge for phenol if they want to make \$0.02/lb.

2. What's wrong with Cutthroat's logic?

ETHYLBENZENE AND STYRENE

Diogenes the wise crept
into his vat
And spoke: "Yes, yes,
this comes from that."

■

Wilhelm Busch Diogenes, 1832–1908
(inventor of the cartoon strip)

Ethylbenzene is to styrene what cumene is to phenol. The only reason you want to make ethylbenzene is so you can make styrene. Its destiny is tied to styrene consumption. Most ethylbenzene (EB) is made by alkylating benzene with ethylene, as shown in Figure 8–1.

A small amount of EB is present in crude oil and also is formed in cat reforming. You might recall from Chapter III that there is only a 4°F difference between the boiling points of EB and para-xylene. Consequently, a super-distillation column is needed for the separation. In process engineers' terms, it would have about 300 theoretical trays, be about 200 feet tall, and even then have a high reflux ratio to accomplish the separation. All this is necessary because the EB stream must be quite pure to be used for styrene manufacture.

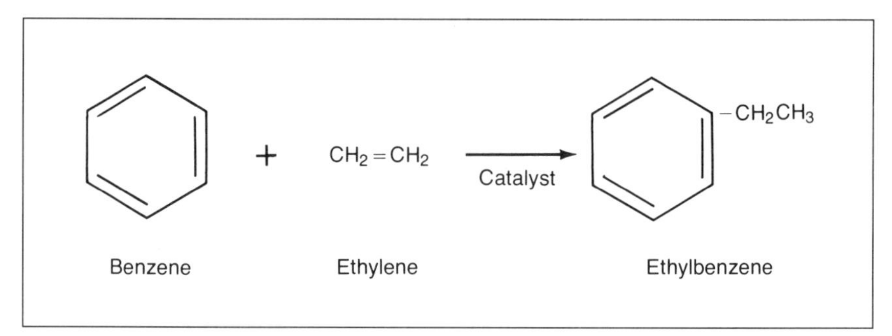

Fig. 8–1 Alkylation of benzene with ethylene to form ethylbenzene

The Technology

Alkylation of benzene is old technology. The French chemist, Charles Friedel, with his American partner, James Crafts, in 1877, stumbled (almost literally) across the technique for alkylating benzene with amyl chloride ($C_5H_{11}Cl$). The use of a metallic catalyst, in this case aluminum, was the key. The Friedel-Crafts reaction is classical and today is the principal route for alkylating benzene with ethylene to make EB.

The Friedel-Crafts reaction has one major drawback. It doesn't stop at the mono-substitution stage. That is, the catalyst works so well, that the benzene will pick up two, three, or more ethylene molecules, forming di-ethylbenzene, tri-ethylbenzene, or higher poly-ethylbenzenes. (See Fig. 8–2.) The problem is that chemically it's easier to alkylate EB than it is benzene. One way to control the problem is to carry out the reaction in the presence of a large excess of benzene. When an ethylene molecule is in the neighborhood of one EB molecule and 20 benzene molecules, chances are that the ethylene will hook up with benzene, even though it prefers EB.

The other controllable variable is the operating conditions. Certain temperature and pressure levels will favor the benzene and not EB alkylation—not exclusively, but predominantly. In fact, these variables can be set to favor the di- and tri-ethylbenzenes to give up an ethyl group to benzene to give EB. That process is called transalkylation and is shown in Figure 8–2.

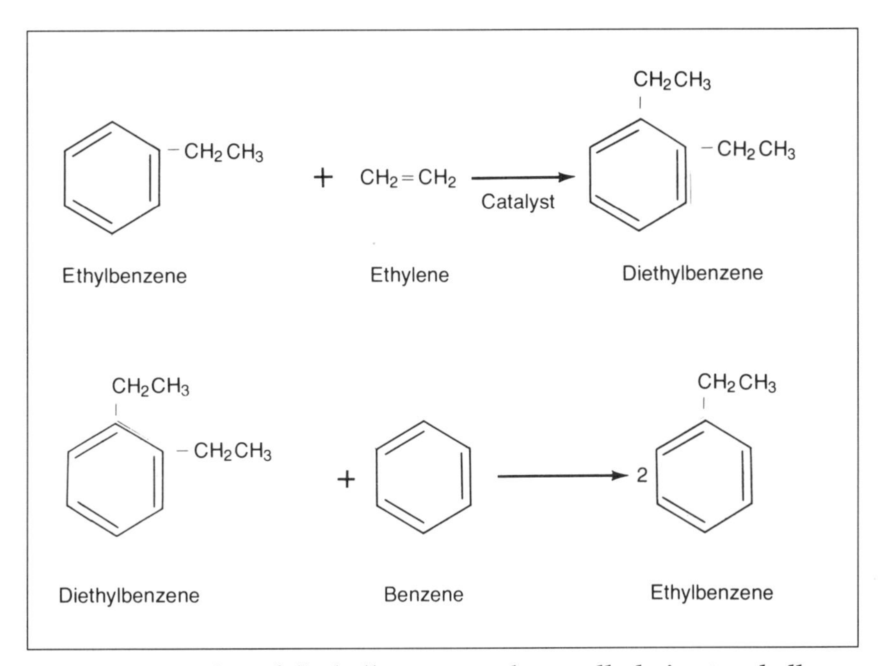

Fig. 8–2 Formation of diethylbenzene and transalkylation to ethylbenzene

The Process

The hardware can be built to accommodate the reaction in either a liquid or vapor phase. The liquid operation is more common and is shown in Figure 8-3. The catalyst used is anhydrous aluminum chloride (anhydrous means completely water free). At temperatures of 300–400°F and pressures of 60–100 psi, the reaction time is about 30 minutes, so the reactor must be large enough to accommodate this long residence time.

Sometimes a catalyst promoter or accelerator, ethyl chloride, is added to the feed to speed up the reaction. The ethyl chloride actually works on the aluminum chloride catalyst, not the reactants. It's like giving a supervisor a bonus. He doesn't do any more work, but he gets more work done.

The effluent stream leaving the reactor is cooled and then treated with caustic (sodium hydroxide) and water to remove the catalyst. The cleaned up stream then contains about 35% unreacted benzene, 50% EB, 15% polyethylbenzene (PEB), and a small amount of miscellaneous heavy

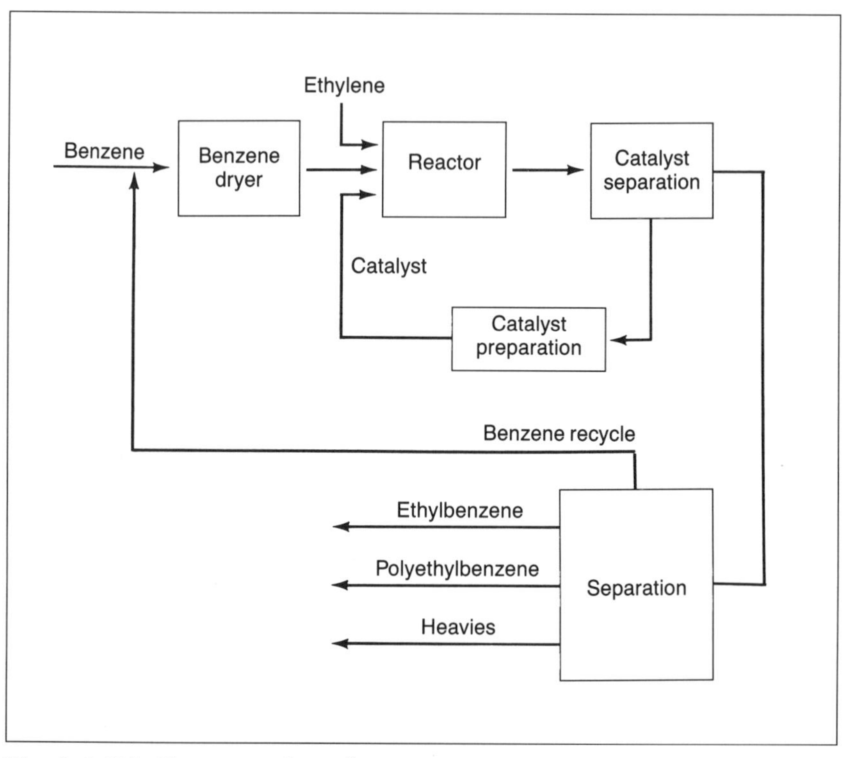

Fig. 8–3 Ethylbenzene plant flows

materials. The EB yields, i.e., the percent of feed that ends up as product, are about 99%, based on the ethylene and benzene feed. (For a discussion of the difference between yield and conversion, see Appendix A.)

The effluent is split into four components in the separation section: unreacted benzene, which is recycled; EB; PEB; and heavier by-products. Usually, the benzene is water washed before it is returned to the feedline; hence, the need for the dryer, which is often an azeotropic distillation column. (See Chapter III.) The presence of any water in the reactor would cause lots of undesirable side reactions to occur.

The PEB is recycled to the reactor also, where transalkylation to EB converts enough of this so that hardly any net PEB is produced. The EB from the second column is about 99% purity.

Material Balance	
Feed:	
Benzene	743 lbs.
Ethylene	267 lbs.
Catalyst	small amount
Ethyl chloride	small amount
Product:	
Ethylbenzene	1000 lbs.
Polyethylbenzenes	10 lbs.

The alternate process, the vapor phase method, is carried out at higher temperatures (485–575°F) and pressures (over 600 psi) using a solid phosphoric acid or boron trifluoride/silica alumina catalyst in a fixed bed. The balance of the process is about the same.

Handling

Since most of the EB is used for the manufacture of styrene, EB plants are usually found in close proximity to styrene plants. Very little EB is traded commercially or transported. A small amount of EB is used as a commercial solvent, mainly as a substitute for xylenes.

EB is not toxic like the xylenes, and it does not require the DOT red shipping label. EB has the same colorless appearance of the other BTX's as well as that characteristic sickly odor. Like the BTX's, it is insoluble in water.

STYRENE

The rapid growth of styrene after World War II was due to the widespread use of its derivatives, principally synthetic rubber and plastics. Styrene ought to be called one of the basic building blocks of the petrochemicals industry. But you get all mixed up with semantics because it's made up of two other basic building blocks, ethylene and benzene. Nonetheless, it is the most important monomer in its class.

You may wonder what makes styrene, an unlikely looking building block, so valuable. There is an explanation from the field of atomic chemistry about the special synergy that results from the combination of the benzene ring and ethylene. Both chemicals are each very reactive. But, while the relationship might intrigue you, the explanation is more complex than can be handled "in non-technical language." In the end, the value of styrene comes from the ease of handling, the safe processing characteristics, and the low costs of the products that can be made from it.

Practically all the styrene today, about 90%, comes from the dehydrogenation of EB—removal of two hydrogen atoms from the ethyl group attached to the benzene ring. (See Fig. 8–4.) The process is similar to operations in an olefins plant (see Figure 8–7)—the dehdrogenation is done by mixing the feed with steam and cracking it in pyrolysis operation. However, the cracking products are more limited, primarily because of the use of a catalyst, iron oxide.

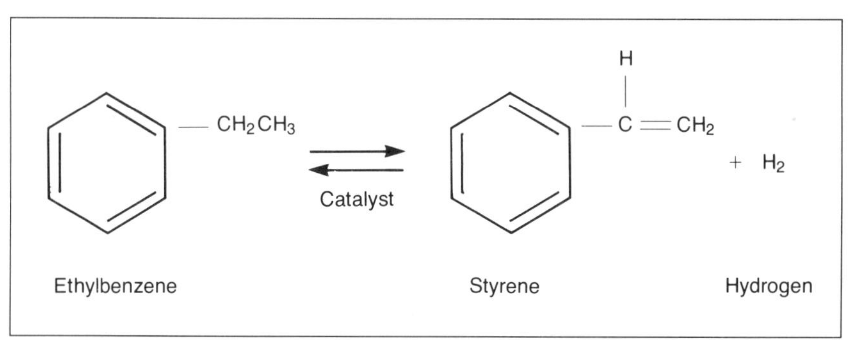

Fig. 8–4 Dehydrogenation of ethylbenzene to styrene

The dehydrogenation step, like all cracking processes, is endothermic (absorbs heat). Superheated steam, mixed with the EB, provides the heat and also performs two other important functions:

1. It reduces the pressure at which the reaction will take place. A mixture of EB and steam at, say, 1300°F can be contained at a lower pressure than EB by itself at that temperature. So what, you say. Well, the chemical reaction of EB cracking to styrene and hydrogen is reversible, Styrene easily hydrogenates back to EB. The reaction which is favored (both will always occur to some extent) will be determined by the operating conditions. In this case, higher pressures favor formation of EB, because EB takes up less volume than the corresponding amount of styrene and hydrogen. Conversely, lower pressures favor formation of styrene. So the logic is that steam mixed with the EB permits cracking the hydrogen off at lower pressures and favors the styrene "staying cracked." (You may have noticed the chemical equation in Figure 8–4 has arrows going both directions. That's the chemist's notation for this reversibility.)

2. The steam also reacts with coke deposits on the iron oxide catalyst, forming CO_2, giving the catalyst a longer, more active lifetime. The on-stream factor of the styrene plant is extended by reducing the shutdown frequency for catalyst regeneration or replacement.

In this process, conversion rate, the percent of feed that "disappears" in one pass through the reactor, is about 60%; the yield, the percent of feed that ends up as product, is about 90%. (Again, the difference between conversion and yield, if you want to know, is discussed in the Appendix.) The 10% yield loss results from cracking one of the carbon-carbon bonds along EB's ethyl group. Consequently, benzene and toluene make up most of the by-products. Another is polystyrene. The styrene, of course, is very reactive. That's why it's a building block. Though the reactor operating conditions minimize this reaction, some of the styrene molecules do join up to form styrene polymers.

In this process, it is very uncommon to detect any benzene rings breaking up. As in the olefins plant, the thermal stability of the benzene ring is demonstrated by its survival under these severe operating conditions, especially the high temperatures.

Material Balance

Feed:	
Ethylbenzene	1132 lbs.
Catalyst	small
Product:	
Styrene	1000 lbs.
Hydrogen	19 lbs.
By-products	113 lbs.

The Process Facilities

The EB entering the styrene plant is generally heated to the threshold cracking temperature (about 1100°F) in a heat exchanger.* The counter flow in the exchanger is the effluent from the second stage reactor, as shown in Figure 8–5. Because high temperatures are necessary in a styrene plant, energy conservation plays a big role in the plant design.

*In a heat exchanger, the two streams trading heat never come in physical contact with each other. One stream is inside a set of pipes, or more commonly, exchanger tubes, that pass through a vessel containing the other stream. The heat exchange comes about by heat (only) passing through the tubes. After all, if the streams were mixed to transfer the heat, they'd have to be reheated to separate them by fractionation.

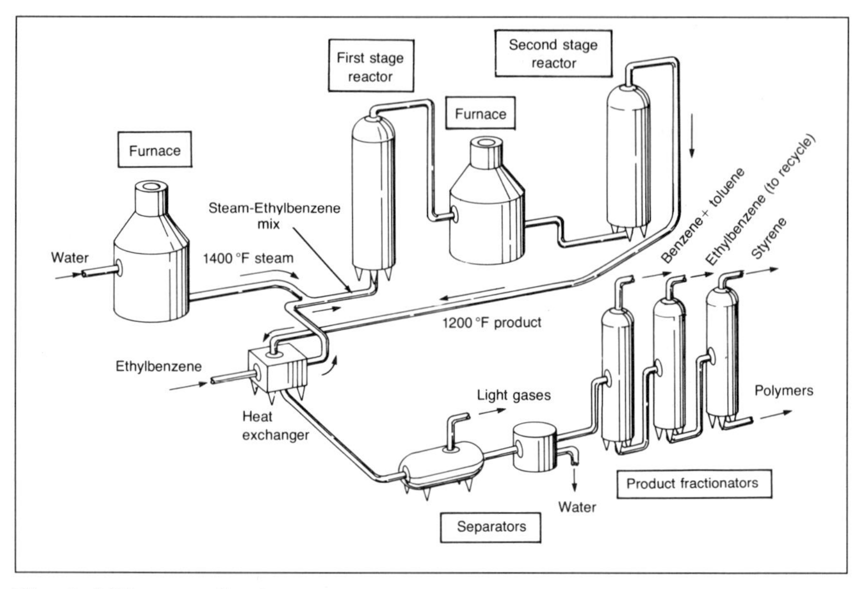

Fig. 8–5 Styrene plant

While this is the only heat exchanger shown, recovery of waste heat is an intimate part of the process flow throughout the plant.

After heating, the EB is mixed with superheated steam and fed to the first stage reactor. Both the first and second stage reactors are packed with a catalyst of metal oxide deposited on an activated charcoal or alumina pellet. Iron oxide, sometimes combined with chromium oxide, or potassium carbonate is commonly used.

The actual reaction takes place at about 1150 °F, but there is a temperature drop in the reactor as the dehydrogenation takes place. Reheating the stream in a furnace or exchanger is necessary before the stream is fed to the second stage reactor for a repeat performance.

The hot effluent is cooled in another heat exchanger, passing the heat off to the incoming fresh EB feed, and maybe even further (not shown) with water to make steam. The cooled stream is then sent to separators where the light cracked gases that are unavoidably formed (H_2, CO, CO_2, CH_4, etc.) and water are removed. The final product separations are done in a series of fractionators. The EB is recycled to the feed line; the polymers, very small in volume, are generally disposed of in residual fuel.

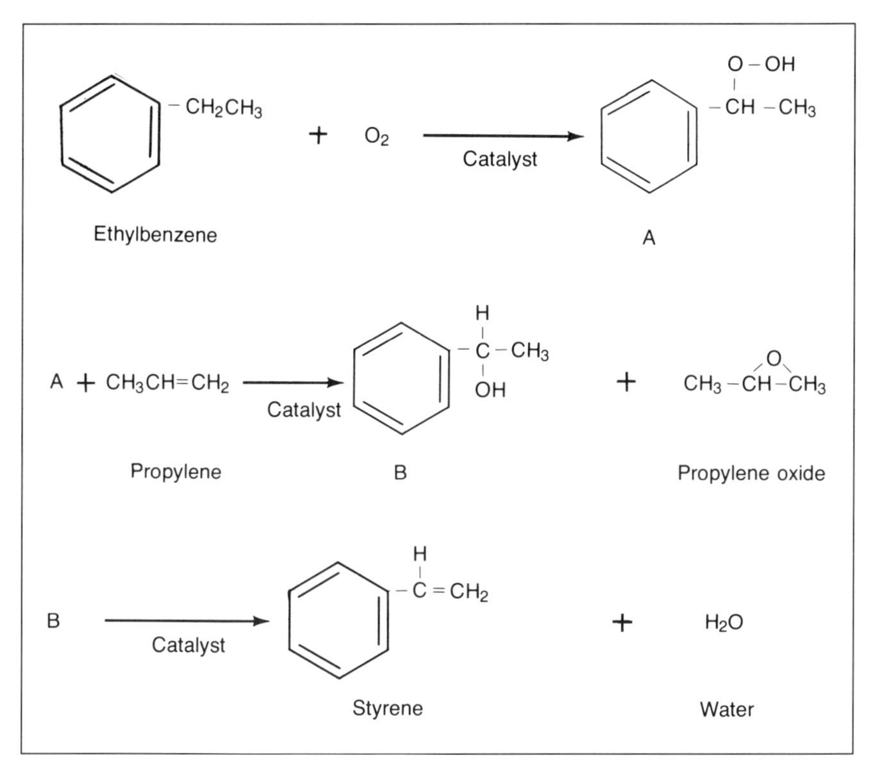

Fig. 8–6 Styrene and propylene oxide by ethylbenzene oxidation

Even at ambient temperatures, styrene is likely to react with itself—very slowly, but steadily. For this reason, a small amount of polymerization inhibitor, about 10 ppm (parts per million) of para-tertiarybutyl catechol, a chemical whose name nobody wants to remember except its salesman, is added to styrene kept in storage. Since polymerization is promoted by higher temperatures, styrene is usually stored in insulated tanks.

Alternate Routes to Styrene

A few plants are designed to produce styrene from EB but as a co-product with propylene oxide. In this process, EB is oxidized to a hydroperoxide (A in Fig. 8–6) by bubbling air through the liquid EB in the presence of a catalyst. Hydroperoxides are, by their nature, very unstable compounds (one of the reasons that bleach, another hydroperoxide, works so well). So exposure to high temperatures has to be limited. The reactions are usually run at about 320°F. Heat exchangers and multiple vessels are used to control the temperatures. Pressures are not critical.

The hydroperoxide is then reacted with propylene at 250°F in the presence of a metal catalyst to produce propylene oxide and methylbenzyl alcohol (B in Fig. 8–6). The latter is easily dehydrated to form styrene. The overall yield of styrene (the amount of EB that ends up as styrene) via this peroxidation process route is 90%.

Commercial Aspects

Uses. Plastics and synthetic rubber are the major uses for styrene. They account for the exponential growth from a few million pounds per year in 1938 to over eight billion pounds today. The numerous plastics include polystyrene, styrenated polyesters, acrylonitrile-butadiene-styrene (ABS), styrene-acrylonitrile (SAN), and styrene-butadiene (SB). Styrene-butadiene rubber (SBR) was a landmark chemical achievement when it was commercialized during World War II. The styrene derivatives are found everywhere—in food-grade film, toys, construction pipe, foam, boats, latex paints, tires, luggage and furniture.

Styrene Properties	
Freezing point	–23.1°F (–30.6°C)
Boiling point	293.4°F (145.2°C)
Specific gravity	0.9045 (lighter than water)
Weight per gallon	7.55 lbs./gal

Handling. Styrene is a colorless liquid but tends toward a yellowish cast as it ages. It feels oily to the touch, and smells like the aromatics compounds. Left alone at room temperature, styrene will eventually polymerize with itself to a clear glassy solid.

Technical grade styrene is 99% minimum purity. It is shipped, with a polymerization inhibitor in it, in standard tank cars or trucks. However, it has none of the severe handling precautions that benzene does.

Fig. 8–7 Polysar, Ltd.'s styrene monomer plant at Sarnia, Ontario.

Chapter VIII in a nutshell...

Ethylbenzene, $C_6H_5C_2H_5$, belongs in the BTX family because it is a benzene ring with an ethyl group, $-CH_2CH_3$, attached in place of a hydrogen. It is made by reacting benzene and ethylene. Virtually all ethylbenzene is used to make styrene.

Styrene is a benzene ring that has a double bonded group attached, which gives it the reactivity that makes it so useful. To convert ethylbenzene to styrene, quick, high temperature exposure in a cracking furnace is used. The ethylbenzene, fortunately, preferentially loses a hydrogen from the ethyl group, leaving a double bonded carbon.

Exercises

1. ____________ is to phenol as EB is to ____________.
2. By-products are important. What are two discussed in the last two chapters that could have important effects on product economics?
3. How many pounds of styrene could you make starting with 1000 pounds of benzene? How much ethylene would you need?

IX

ETHYLENE DICHLORIDE AND VINYL CHLORIDE

"Inventing is a combination
of brains and material.
The more brains you use,
the less material you need."

■

**Charles F. Kettering, 1876–1953,
President, General Motors**

This is the third chapter in a series of three where the products are like pancakes and batter. You can't make pancakes (or vinyl chloride) without making batter (or ethylene dichloride); there's not much else you can do with pancake batter (or ethylene dichloride); and you only use pancakes or vinyl chloride to make something else—breakfast or plastics. Finally, if you'll permit this overbearing analogy to be extended once more, making vinyl chloride (or pancakes) from scratch is a lot easier now than it was 40 years ago.

The original manufacturing route to vinyl chloride (VC) didn't involve ethylene dichloride (EDC) but was the reaction of acetylene with hydrochloric acid. This process was commercialized in the 1940s, but like most acetylene-based chemistry in the U.S., it gave way to ethylene in the 1950s and 1960s. The highly reactive acetylene molecule was more sensitive, hazardous, and eventually more costly than ethylene. The chemical engineers were happy to replace acetylene technology with the ethylene route. All the contemporary vinyl chloride plants now use ethylene and chlorine as raw materials.

Vinyl chloride is often called vinyl chloride monomer (VCM). The tag-on, monomer (from the Greek mono, meaning one and meros, meaning part) is a convention used to contrast a chemical from its counterpart, the polymer. Vinyl is the prefix for any compound that has the vinyl group, $CH_2=CH$-, in it. The route of the word is the Latin vinum, meaning wine, perhaps having something to do with a preoccupation of the discoverer.

THE PROCESS

VC is made by cracking EDC in a pyrolysis furnace much like that in an ethylene plant. That's one of the three reactions, shown in Figure 9–1, involved in the process. The other two have formidable names—chlorination and oxychlorination—but simple enough reactions—the addition of chlorine and the addition of oxygen and chlorine. What is a little complicated is the fact that the hydrogen chloride used to make the EDC in the first reaction comes from cracking EDC in the second. Sounds like a closed circle until you peel it back and examine it.

The plant, with its three reactors, is shown in Figure 9–2. One of the reactors is the pyrolysis furnace in the middle of the figure. At the top of the figure, the basic feeds to the plant are shown—ethylene, chlorine, and oxygen. Ethylene and chlorine alone are sufficient to make EDC via the route on the left. The operation, call it Reaction One, takes place in the vapor phase in a reactor with a fixed catalyst bed of ferris (iron) chloride at only 100–125 °F. A clean-up column fractionates out the small amount of by-products that get formed, leaving an EDC stream of 96–98% purity.

For Reaction Two, the purified EDC is passed through a dryer to remove water, then fed to a pyrolysis unit. The difference between EDC pyrolysis furnaces and those used for ethylene is the use of a catalyst.

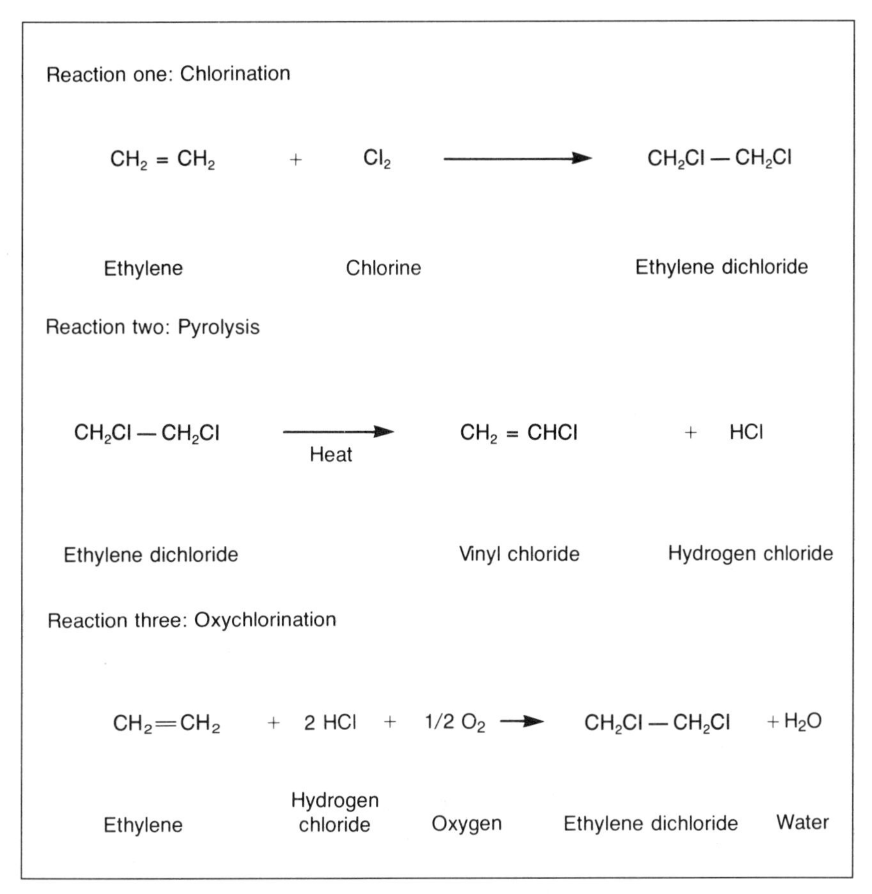

Fig. 9–1 Ethylene dichloride and vinyl chloride reactions

The tubes in the EDC furnace are packed with charcoal pellets impregnated with ferric (iron) chloride. The EDC is pumped through at about 900–950°F and 50 psi. The conversion of EDC, i.e., how much of it "disappears," is about 50% and the yield of VC, how much of the "disappearing" EDC gets converted to VC, is about 95–96 percent.* So not much else is formed. That's a contrast to ethylene manufacture, especially cracking the heavy liquids, where the by-products are abundant.

The hot effluent gas from the furnaces is quenched right away for the same reasons as ethylene furnace effluent is quenched—to stop the

*See the Appendix if you haven't yet read about the difference between yield and conversion.

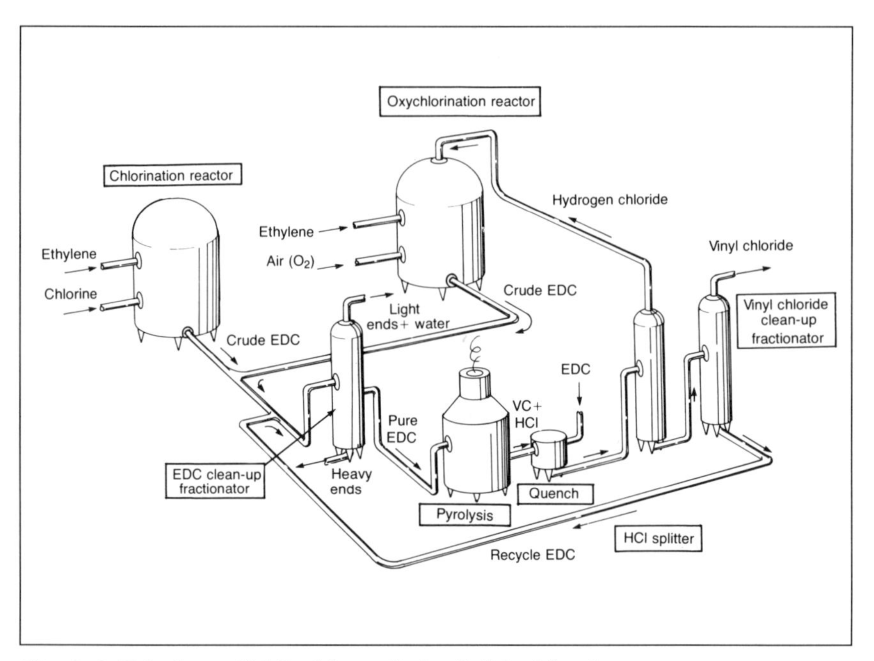

Fig. 9–2 Ethylene dichloride and vinyl chloride plant

cracking at the optimum point. In this case, though, the quench liquid is cool EDC, not water.

When EDC cracks, one hydrogen and one chlorine on adjacent carbon atoms are sprung and find each other, forming hydrogen chloride gas. (Why one of each cracks off, and not two hydrogens or two chlorines is another mystery of atomic physics.) The cooled effluent is fractionated into three streams: hydrogen chloride, EDC, and the VC stream which is sent to storage. The EDC, which is the unconverted pyrolysis feed plus what was added in the quench pot, is recycled to the EDC clean-up column. The hydrogen chloride, which would otherwise be a disposal problem, is pumped to the Oxychlorination Reactor, as shown in the upper right corner of Figure 9–2.

The oxychlorination reactor is packed with cupric (copper) chloride catalyst. Three feeds, gaseous hydrogen chloride, pure oxygen or oxygen in the form of air, and ethylene are reacted at 600–800 °F, to form EDC and water. The reaction effluent is then piped over to the other EDC stream heading for the clean-up fractionator, where it comingles with the EDC stream from Reaction One and the recycle stream from the VC fractionator.

Material Balances	
Feed:	
Ethylene	295 lbs.
Chlorine	750 lbs.
Product:	
Ethylene dichloride	1000 lbs.
By-products	45 lbs.
Feed:	
Ethylene dichloride	1667 lbs.
Product:	
Vinyl chloride	1000 lbs.
Hydrogen chloride	578 lbs.
By-products	89 lbs.

So, there are two recycle streams, hydrogen chloride and EDC. The EDC is recycled to pyrolysis; the hydrogen chloride is recycled to the oxychlorination to form EDC. Considerable attention has to be paid to balancing the flows around this plant. There are surge tanks in the plant that are not shown in Figure 9–2. But they can quickly fill up, potentially causing the need to shut down one of the reactions to play catch-up. Starting and stopping any of the reactions tends to be a problem, both in off-spec product and wasted energy costs.

HANDLING CHARACTERISTICS

Sufficient evidence has proven that VC can cause cancer of the liver after prolonged exposure to only minute quantities (parts per million). Elaborate hardware precautions are taken to eliminate escape of any VC to the atmosphere. Personnel involved in production or use of VC often wear respirators whenever there is the possibility of a leak.

VC vaporizes at about 7°F, so at normal temperatures it must be contained in pressure vessels to keep it liquid. This includes movement by tank cars and trucks, which must fly the hazardous material sticker enroute.

Properties	
EDC:	
Freezing point	–31.7°F (–35.4°C)
Boiling point	182.3°F (83.5°C)
Specific gravity	1.253 (heavier than water)
Weight per gallon	10.5 lbs. per gal.
VC:	
Freezing point	–244.8°F (–153.8°C)
Boiling point	7.9°F (–13.37°C)
Specific gravity	0.9106 (lighter than water)
Weight per gallon	8.14 lbs./gal.

VC is highly reactive, and like styrene, will start to polymerize with itself if it just sits in a tank. Phenol, in trace amounts, is an effective polymerization inhibitor and is normally added to VC on the way to storage.

EDC is a much less nasty commodity. It need not be shipped in a pressurized vessel, but it is classified as a hazardous material and must be kept in a closed system.

COMMERCIAL ASPECTS

Nearly all VC is used to manufacture polyvinyl chloride in a polymerization process described about 10 chapters from here. Most EDC is used to manufacture VC, but some is used as a lead scavenger (cleaner-upper) in gasoline; in the manufacture of perchloroethylene, an industrial degreaser and dry cleaning agent; and in the manufacture of methyl chloroform, an anesthesia.

EDC and VC are each traded commercially as a 99% purity grade. VC is usually designated as inhibited, indicating the presence of phenol.

Chapter IX in a nutshell...

Ethylene dichloride, $C_2H_4Cl_2$, is a petrochemical that is made so that vinyl chloride can be made out of it. The process for making EDC is sometimes integrated with the vinyl chloride plant. EDC is made by reacting ethylene with hydrogen chloride.

Vinyl chloride, C_2H_3Cl or $CH_2{=}CHCl$, is ethylene with a chlorine atom replacing a hydrogen. It is made in two ways. EDC is subjected to high temperatures in a cracking furnace, where a chlorine and a hydrogen atom pop off, leaving vinyl chloride. The availability of that free chlorine atom makes it appropriate to make hydrogen chloride, giving rise to the other route, where the hydrogen chloride, ethylene, and oxygen are reacted to make vinyl chloride and water.

Almost all vinyl chloride is used to make polyvinyl chloride, a versatile consumer plastic.

Exercises

1. Mix and match:

quench	HCL, O_2, and C_2H_4 to EDC
oxychlorination	hot VC/HCL plus cool EDC
pyrolysis	C_2H_4, Cl_2, to EDC
chlorination	EDC to VC and HCl

2. Take the three reactions for the EDC/VC plant and "net them out" algebraically, showing how much goes in the plant, how much comes out. (Hint: The only products are VC and water.)
3. What's the difference between chlorination, oxidation, and oxychlorination?

ETHYLENE OXIDE AND ETHYLENE GLYCOL

"From out of the past come the thundering hoofs of the giant horse Silver."

■

from *The Lone Ranger Rides Again*
Fran Striker

In the preceding chapters, all the petrochemicals discussed and their immediate derivatives had double-bonded carbons imbedded in their structure. This characteristic makes those chemicals very reactive, which is why they are so useful as building blocks. In contrast, ethylene oxide, EO, has no double bond, but instead a three member heterocyclic ring, with "hetero" meaning one of the atoms isn't carbon—it's oxygen. (See Fig. 10–1.) This cyclic oxide is often called an epoxide. The suffix ep- is from the Latin meaning "on" or "beside." In chemistry, ep- generally refers to the heterocyclic ring. The other common chemical with this suffix is epichlorohydrin.

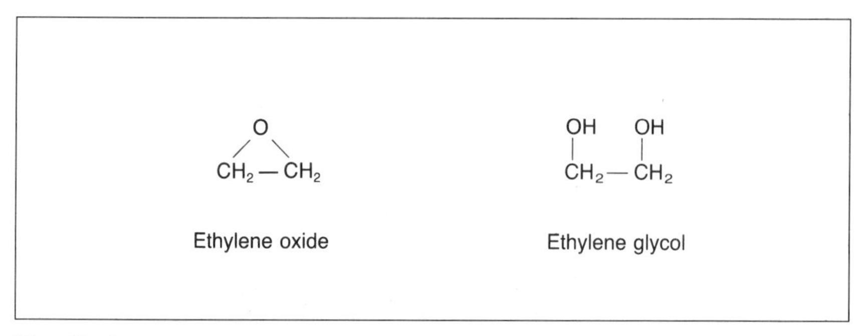

Fig. 10–1

When EO is formed, single bonds from two adjacent carbons are connected to an oxygen atom. A three member ring is always in a "strained" condition, due to the geometry of the molecule. Because of the propensity to relieve the strain, epoxides are very reactive. Practically all the EO produced is converted to chemical intermediates as a result of a ring opening reaction.

The key feature of ethylene glycol, EG, is the hydroxyl group, -OH, one on each of the two carbon atoms. The hydroxyls are responsible for its reactivity: EG is a monomer used in the production of polyester polymers. The hydroxyls also give EG its most important physical property, its solubility in water. That, linked with its low freeze point, makes EG suitable as antifreeze and as a de-icer. When EG is sprayed on ice, it combines with the water crystals and lowers the freeze point. This causes the mixture to melt and effectively keeps it in the liquid state.

ETHYLENE OXIDE

Until the 1940s, the commercial route to EO was ethylene chlorohydrin, $\overset{OH}{\overset{|}{CH_2}}\text{-}\overset{Cl}{\overset{|}{CH_2}}$. This was a two step process—conversion of ethylene to the chlorohydrin by reaction with hyporchlorous acid, HO-Cl, followed by dehydrochlorination (removal of the HCl) of the ethylene chlorohydrin to give EO.

The problem with the process was not the yield of EO but the operating expenses, particularly the cost of chlorine. Almost all the chlorine introduced as part of the HO-Cl ended up after the process as calcium chloride. Not only was this compound a worthless solid, it had to be hauled away for disposal.

In the 1940s and 1950s, a considerable amount of research was funded to find and develop the chemist's impossible dream, a process for the direct oxidation of ethylene to EO, without any by-products. Finally, Union Carbide found the silver bullet that did the job—a catalyst made of silver oxide. The relationship between this catalyst and the reaction is unique. Silver oxide is the only substance found having sufficient activity and selectivity. (Activity relates to the amount of conversion, selectivity relates to the right yield.) Moreover, ethylene is the only olefin affected in this way. The others, propylene, butylene, etc., tend to oxidize completely, forming carbon dioxide and water. But when silver oxide is used as a catalyst with ethylene, the dominant reaction is the formation of EO. Some ethylene still ends up being further oxidized, as much as 25% in some processes, as shown in Figure 10–2.

The process was commercially so superior to the chlorohydrin route, that by the 1970s, the new chemistry had completely replaced the old. Adding some momentum to this transition was the fact that the obsoleted and abandoned chlorohydrin plants could be readily converted to propylene oxide plants. The silver bullet for that process has yet to be found.

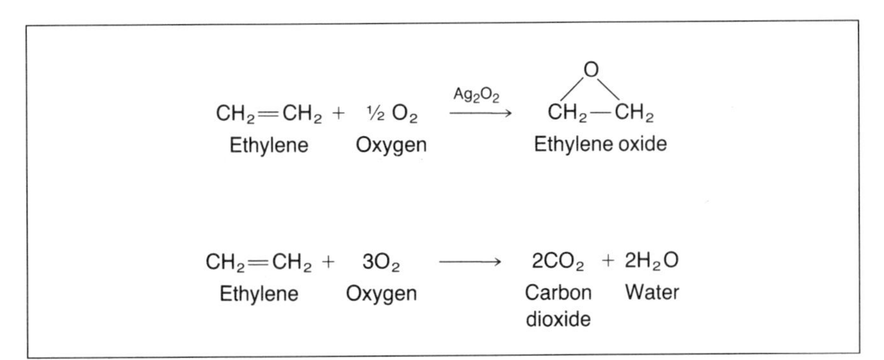

Fig. 10–2 Direct oxidation of ethylene to ethylene oxide

The Process and The Hardware

The new EO plants are as simple as any you will read about in this book. The feeds are mixed, reacted, then split into recycle and finished product streams, as shown in Figure 10–3. The oxidation reaction takes place in the vapor phase. Ethylene and oxygen, either in the form of air, pure oxygen, or air enriched with pure oxygen, are heated, mixed, and then passed through a reactor with fixed beds of catalyst—silver oxide deposited on alumina pellets. The reaction takes place at 500–550 °F under slight pressure. The residence time of the feed in the reactor is only about one second. Yields, the amount of the ethylene that ends up as EO, are in the 70–75% range.

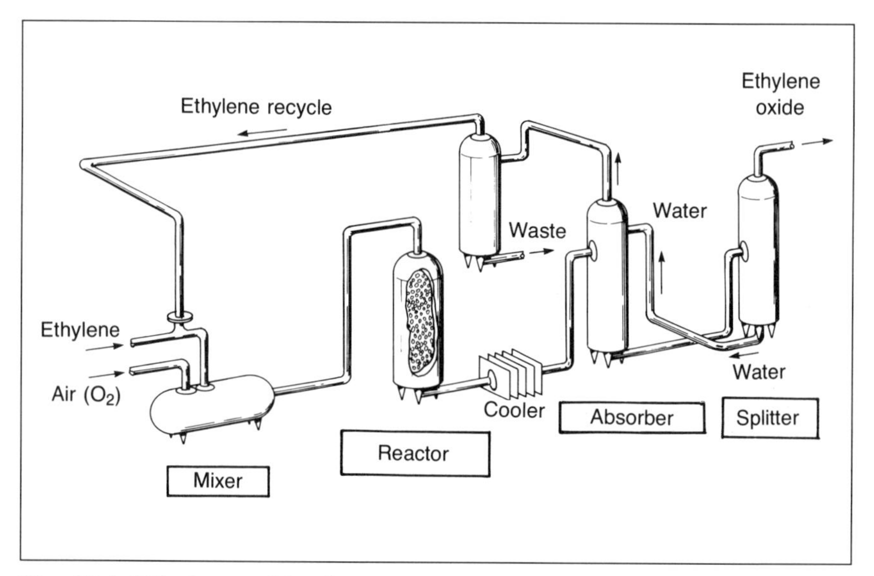

Fig. 10–3 Ethylene oxide plant

Like most oxidations, this one is exothermic. The effluent from the reactor is cooled in a heat exchanger. The EO, by-products, and unreacted ethylene are separated in a water-wash column in a manner just like the solvent recovery process described in Chapter II. The EO is absorbed by the water, the by-products (mainly CO_2, plus the ever-present cats and dogs in small quantities) and unreacted ethylene are not. The waste water/EO are easily split in the next column.

The by-products and the ethylene are split, and the ethylene is recycled to the reactor. (The by-product splitter is not drawn correctly in Figure 10–3, because some of the by-products actually have lower boiling points than ethylene. The by-product splitter should really be shown as a series of columns.)

The early but new technology plants in the 1950s used air as the source of oxygen. In 1956, the pure oxygen technique was first introduced in the U.S. Which is better is still being debated. As a result, about 60% of the presently produced EO is based on air.

Various marginal improvements have been introduce over the years, like the used of catalyst promoters and catalyst inhibitors which suppress the tendency for CO_2 formation.

Material Balance	
Feed:	
Ethylene	850 lbs.
Excess air	11,659 lbs. (2331 lbs. O_2)
Product:	
Ethylene oxide	1,000 lbs.
Carbon dioxide	669 lbs.
Water	274 lbs.
Unreacted air	10,566 lbs. (1227 lbs. O_2)

Commercial Aspects

EO is a colorless gas at room temperature. It boils at 56°F. As a liquid it is colorless, high flammable, explosive and is very soluble in water and common solvents. EO is a toxic substance requiring care in handling.

EO is traded commercially as a high purity technical grade, 99.7% purity. Because of its low boiling temperature, EO must be stored and shipped in vessels that can withstand mild pressures. Trucks and tank cars must fly the red hazardous material label.

Properties	
Freezing point	−169.0°F (−112.0°C)
Boiling point	56.3°F (13.5°C)
Specific gravity	0.8969 (lighter than water)
Weight per gallon	7.45 lbs./gal.

EO is an intermediary chemical and has multiple end-uses. The predominant derivative is ethylene glycol, which uses up more than 60%

the total EO. Biodegradable detergents and ethanolamine are the next major outlets for EO. Others include fumigants (one of the oldest uses), sterilizing agents, latex paints, explosives, fibers, soaps, and pharmaceuticals.

ETHYLENE GLYCOL

The most widely used process for making EG is shown in Figure 10–4. It's even simpler than the EO plant. The EO ring is easily opened in the presence of water with trace amounts of sulfuric acid. So the processing scheme requires a mixing vessel where the following reaction can take place:

$$\underset{\text{(ring, O bridging)}}{CH_2-CH_2} + H_2O \xrightarrow[\text{catalyst}]{\text{Acid}} \underset{OH\ \ \ OH}{CH_2-CH_2}$$

The other three pieces of hardware in Figure 10–4 are used to recycle the EO and to clean up the EG by splitting out the heavier glycol by-products.

In the reactor, the operating conditions to give 95% yields are slight pressure, 125–160°F, and a ratio of water to EO of 10 to 1. The reaction time is slow. This means that the reactor must be large enough so that high throughput rates still allow the molecules a long enough residence time—about 30 minutes.

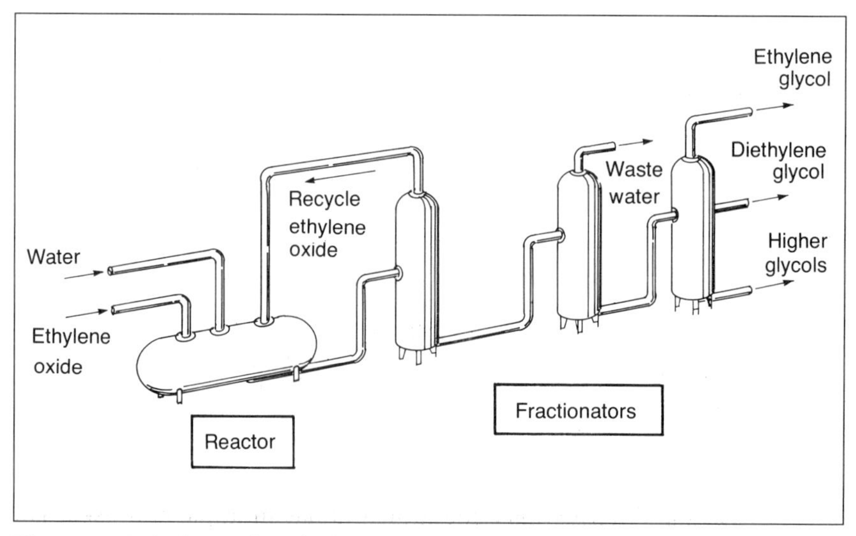

Fig. 10–4 Ethylene glycol plant

Material Balance

Feed:	
Ethylene oxide	816 lbs.
Water (excess)	3263 lbs.
Sulfuric acid	trace
Product:	
Ethylene glycol	1000 lbs.
Diethylene glycol	100 lbs.
Heavy glycols*	30 lbs.
Unreacted water	2949 lbs.

*tri- and tetraethylene glycols, polyethylene glycols, etc.

Since more than 60% of the EO production is converted directly to EG, the obvious question some macho chemist might ask is "why don't we do an end run and just convert ethylene directly to EG? Skip the oxidation step." Research in the 1950s and 1960s led to several promising commercial processes, oxychlorination and acetoxylation. Exotic catalysts were used and both avoided the EO step. But neither process was quite effective enough to replace the ethylene-to-EO-to-EG route which predominates today.

Commercial Aspects

End Uses. It's a little curious that the two major end uses for EG are so different. One is a consumer product, the other is a feedstock for more complicated chemistry. The reasons have to do with two separate properties of EG, one physical property, one chemical property. Because of EG's low freezing point, it is the main ingredient in automotive antifreeze. Because it is so chemically reactive, it is used as a monomer in making polyester polymers. The polymers are used in fiber, film, plastics, and blow-molded bottles. Together, antifreeze and polyester polymers account for about 85% of the ethylene gylcol produced in the U.S. It also is also used sometimes as a deicer for aircraft windshields. The two hydroxyl groups in the EG molecule also make EG suitable for the manufacture of surfactants and in latex paints.

Ethylene Glycol Properties

Freezing point	11.3°F (–11.5°C)
Boiling point	387.7°F (197.6°C)
Specific gravity	1.1108 (heavier than water)
Weight per gallon	9.3 lbs./gal.

Properties and handling. Ethylene glycol is a clear, colorless, syrupy, and virtually odorless liquid. It is hydroscopic, i.e., it absorbs water readily. And, when added to water, it lowers the freeze point. It is traded commercially as a high purity technical grade at 99% content. Ethylene glycol is a friendly liquid, and no particular precautions need to be taken in transporting it by barge, tank car, or truck, as long as you don't fall in and lower your freeze point.

Chapter X in a nutshell...

Ethylene oxide is a triangle-shaped cyclic compound. The "tightness" or shape of the bonds connecting the oxygen and the two CH_2 groups gives rise to EO's chemical reactivity. It readily converts to ethylene glycol in the presence of water and a little acid. It also is used to make polymers and germicides. EO is made by direct oxidation of ethylene, which is greatly facilitated by the unusually effective catalytic power of silver oxide.

Ethylene glycol, CH_2OHCH_2OH, looks like ethylene with two hydroxyl groups, -OH, on each carbon in place of a hydrogen. EG is used as the essential ingredient in antifreeze and in the production of polyester film, fiber, and plastics.

Exercises

1. Mix and match:

epoxide	EG
silver bullet	heavy glycols
epoxide ring	EO
antifreeze	CO_2
EO by-product	silver oxide, the EO catalyst
EG by-product	a molecule under stress

2. What caused the demise of the chlorohydrin route to EO?
3. The Clondike Coolant Company expects to sell 10 million gallons of ethylene glycol to antifreeze blenders this summer. They'll add some additives (maybe), some blue or red or yellow dye, and maybe some water and mark it up about 400%. How much ethylene should Clondike contract for, theoretically, to feed to their ethylene oxide/ethylene glycol plants this spring?

XI

PROPYLENE OXIDE AND PROPYLENE GLYCOL

"Something old, something new,
something borrowed . . ."

▪

Anonymous, *Wedding Rhyme*

You have to talk about propylene oxide and propylene glycol after ethylene oxide and glycol. It's not that the chemical configurations are so similar (they are), or that the process chemistry is about the same (it is). The fact is that much of the propylene oxide is now made in plants originally designed and constructed to produce EO, not PO. As you read in the last chapter, the chlorohydrin route to EO was abandoned by the 1970s in favor of direct oxidation. At the same time, the EO producers found that the old EO plants were suitable for the production of PO, and certainly the cheapest hardware readily available to satisfy growing PO demands.

PROPYLENE OXIDE

The chemical structure of PO differs from EO by the methyl group (-CH_3), as shown in Figure 11–1. The difference is more than just a matter of geometric symmetry. The methyl group in PO, for example, increases the reactivity of the molecule in an adverse way. In the various manufacturing routes to PO, consequently, the reaction of propylene with chlorine or oxygen is hard to stop. By-products are an unavoidable nuisance, and the yields of PO are not as high as the chemical engineers would like.

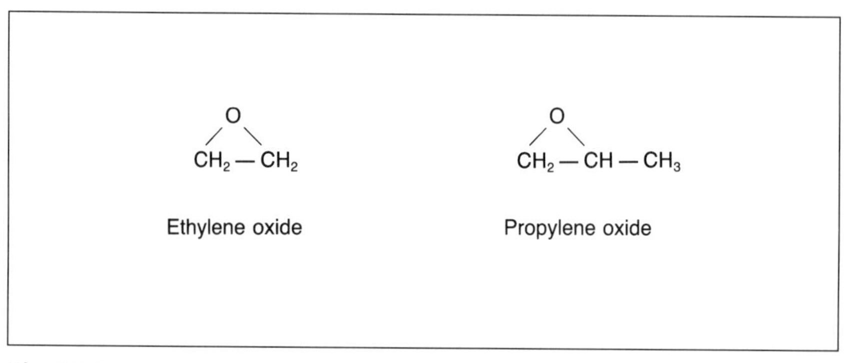

Fig. 11–1

Prior to the late 1970s, almost all PO was produced using the chlorohydrin route, much of it in the former EO plants. But this process was encumbered with the same problems as the EO—it was energy intensive, by-product yield was too high, and the chlorine waste product was a disposal expense. As a consequence, newer technology emerged in the 1980s—the indirect oxidation route. It involves the oxidation of a hydrocarbon (call it R-H) to form a hydroperoxide (with the signature, R-OOH) which is then reacted with propylene to form PO and an alcohol co-product (having the signature R-OH). The market value of the alcohol assists materially in justifying the economics of this alternate route.

The commercial success of the indirect oxidation route has dampened enthusiasm for continuing the search for the "silver bullet" catalyst that will facilitate direct oxidation, *a la* EO. Further activity in that area seems more and more like the quest for that Philosopher's Stone that turns lead into gold.

The Chlorohydrin Route

The chlorohydrin route takes two steps. The description of them, unfortunately, is a sentence whose average word length is nine letters: reaction of propylene with hypochlorous acid (HO-Cl) followed by dehydrochlorination of the propylene chlorohydrin with calcium hydroxide. That's a tough way of saying that a chlorine atom (Cl-) and a hydroxyl group (-OH) are added to the propylene double bond, and then a chlorine atom and a hydrogen atom are removed, leaving the oxygen bonded to two adjacent carbon atoms to form propylene oxide.

Three equations describe the process. The first involves making the hypochlorous acid by reacting chlorine and water. In the second, the acid reacts with propylene to make the chlorohydrin. The dehydrochlorination takes place in the third to give propylene oxide.

Hypochlorous Acid

$$Cl_2 + H_2O \longrightarrow HOCl + HCl$$

Formation of Propylene Chlorohydrin

$$CH_3\text{-}CH{=}CH_2 + HOCl \longrightarrow CH_3\text{-}\underset{|}{\overset{OH}{C}}H\text{-}\overset{Cl}{\underset{|}{C}}H_2$$

Dehydrochlorination to PO

$$2CH_3\text{-}\overset{OH}{C}H\text{-}\overset{Cl}{C}H_2 + Ca(OH)_2 \longrightarrow 2CH_3\text{-}CH\text{-}CH_2\ (\text{with } O \text{ bridging the two carbons}) + CaCl_2 + 2H_2O$$

One other reaction not shown is the formation of propylene dichloride. The demand for this compound is generally insufficient to absorb all the co-production, so it also ends up on the list of "things to be disposed of coming from the PO-chlorohydrin process." But despite this and all the other problems already mentioned about the chlorohydrin route, the process remains economically healthy—breathing heavily, but healthy. Indeed, 40 to 50% of the PO produced in the U.S. comes from this route.

The Chlorohydrin Hardware

Two of the reactions take place in the same reactor in this plant. The formation of the hypochlorous acid (HOCl) from chlorine and water and the reaction with propylene all occur simultaneously on the left in Figure 11-2. Propylene reacts readily with chlorine to form that unwanted by-product, propylene dichloride. To limit that, the HOCl and HCl is kept very dilute. But as a consequence, the concentration of the propylene chlorohydrin leaving the reactor is very low—only 3–5%! At any higher concentration, a separate phase or second layer in the reactor would form. It would preferentially suck up (dissolve) the propylene and chlorine coming in, leading to runaway dichloride yields. The low concentration levels of the propylene chlorohydrin and the need to recycle so many pounds of material is the reason the process is so energy intensive. It just takes a lot of electricity to pump all that stuff around.

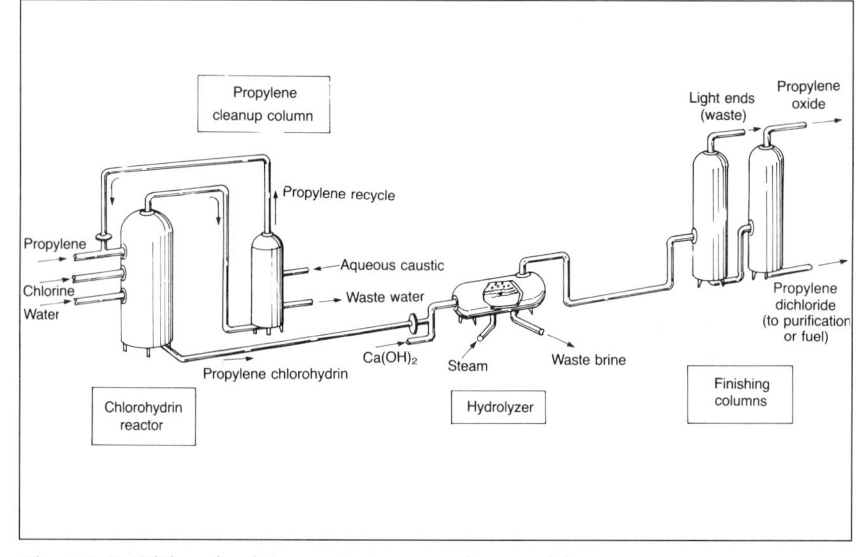

Fig. 11–2 Chlorohydrin route to propylene oxide

The unreacted propylene is taken off the top of the reactor and cleaned up for recycling. By bubbling this stream through a dilute caustic solution (like sodium hydroxide, NaOH), the chlorine and HC1 carried along with the propylene are removed by converting them to sodium chloride, NaCl, and water. The "scrubbed" propylene is then taken overhead (from the top of the fractionation column) and is ready as fresh feed or use elsewhere in the plant.

The dilute propylene chlorohydrin stream is mixed with a solution of water and 10% slaked lime, calcium hydroxide, and pumped to a vessel called the hydrolyzer. The chlorohydrin rapidly dehydrates to PO. The reaction is so fast that the PO has to be sprung from the mixture before the reaction continues, forming propylene glycol. Steam is bubbled through the reactor, helping to flash (vaporize) the PO out of the reaction zone.

The vapor from the hydrolyzer contains not only water and PO, but also propylene dichloride and whatever other cats and dogs (by-products). Fractionation columns are used to purify the PO to a 99% technical grade.

Material Balance
(PO via the chlorohydrin route)

Feed:	
Propylene	941 lbs.
Chlorine	1590 lbs.
Slaked lime $(Ca(OH))_2$	636 lbs.
Products:	
Propylene oxide	1000 lbs.
Calcium chloride	955 lbs.
Hydrogen chloride	628 lbs.
Propylene dichloride	437 lbs.
By-products	147 lbs.

The Indirect Oxidation Route

This indirect oxidation route takes two steps. In the first, a hydrocarbon, such as iso-butane or ethylbenzene, is oxidized. The source of the oxygen is air. The reaction takes place just by mixing the ingredients and heating them to 250–300 °F at 50 psi, producing a hydroperoxide. In the second step, the oxidized hydrocarbon reacts with propylene in a liquid phase and in the presence of a metal catalyst at 175–225 °F and 550 psi to produce PO yields of better than 90%. The process flow is shown in Figure 11-3.

The reaction sequence is summarized below using iso-butane as the hydrocarbon. The crucial and nearly incredible part of the process is in two parts itself, but only shows as one. It is the second equation where the oxygen molecule transfers to the propylene molecule and the ring closes to form the epoxide. (That's why they call it epoxidation.) The magic that causes all that to happen is in the metal catalyst. This is one of many examples in chemistry where catalysts can cause atoms to slide around molecules in unlikely ways.

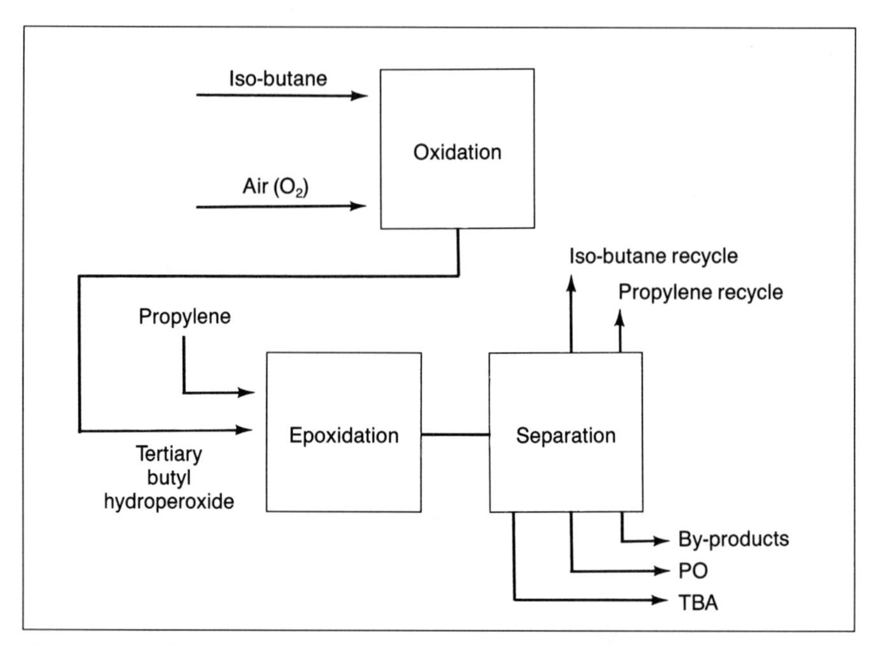

Fig. 11–3 Indirect oxidation route to PO (and TBA)

Hydroperoxide formation (oxidation of is-butane)

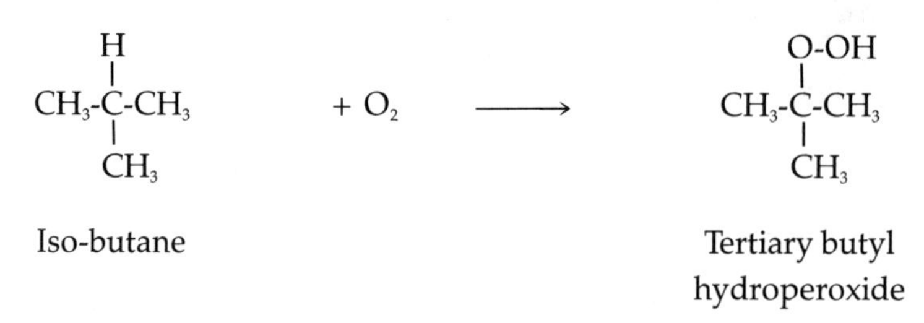

Epoxidation of Propylene (reaction with propylene)

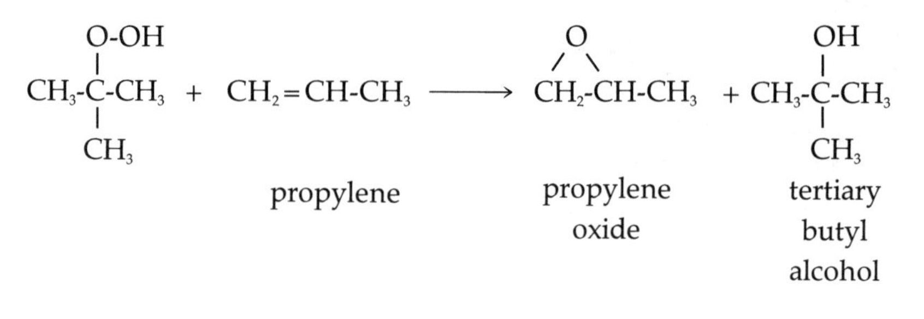

One of the commercial benefits of this route is the value of the co-product, tertiary butyl alcohol (TBA) when iso-butane is used, styrene when ethylbenzene is used. The commercial applications and value of styrene were discussed in Chapter VIII. As for TBA, in Chapter XII its use as a precursor for MTBE will be treated. TBA also can be easily hydrotreated back to iso-butane if a recycle stream for PO manufacture is more advantageous.

Material Balance
(PO Via the Indirect Oxidation Route)

Feed:	
Propylene	782 lbs.
Iso-butane	1880 lbs.
Oxygen (in excess)	1069 lbs.
Catalyst	small
Product:	
Propylene oxide	1000 lbs.
Tertiary butyl alcohol	2400 lbs.
Unreacted oxygen	269 lbs.
By-products	62 lbs.

Commercial Aspects

Although propylene oxide is structurally similar to ethylene oxide, its applications are very different. For example, propylene glycol accounts for only about 20% of PO demand, compared to the 60–65% share EG takes of the EO. About 60% of the PO is used to make polyethers and polyester polyols. These are chemical derivatives that are reacted with diisocyanates to make flexible and rigid polyurethane foams. You're probably sitting on some now (flexible, not rigid, if you're lucky).

Propylene glycols and dipropylene glycols are used to make thermoset polyester resins for use in fiberglass composites. Fabricated products include boat hulls, shower stalls, appliance casings, furniture, and automobile parts. The polyester is usually reinforced with shredded, chopped, or woven glass fiber. Ironically, even though boats are said to have fiberglass hulls, most of the material by weight is polyester made from propylene oxide.

Smaller but growing markets for PO include liquid detergents and surface coatings. The PO route to butanediol is replacing the acetylene route.

Propylene Oxide Properties	
Freezing point	–155.2°F (–104.0°C)
Boiling point	93.6°F (34.2°C)
Specific gravity	0.826 (heavier than water)
Weight per gallon	6.92 lbs./gal.

Propylene oxide is a low boiling, flammable liquid, readily soluble in both water and the more common organic solvents, such as alcohol, ether, aliphatic and aromatic hydrocarbons. Commercial sales involve only technical grade (about 98%), and bulk movements require a hazardous material shipping label. Standard transport equipment (trucks, tank cars, and barges) can be used.

PROPYLENE GLYCOL

This has to be the quickest product treatment in this book—if you've already read the ethylene glycol chapter. The process for propylene glycol is the same as for EG. A little sulfuric acid in water at about 150°F will open up the epoxide ring, and the water will provide the hydroxyl groups to form propylene glycol. With plenty of excess water, high yields of propylene glycol are achieved. However, some higher glycols, primarily dipropylene glycol, will show up as by-products.

```
        O
       / \                         dilute           OH  OH
  CH3-CH -CH2          + H2O      -------->          |   |
                                    acid         CH3-CH -CH2
```

$$CH_3\text{-}\overset{O}{CH\text{-}CH_2} + H_2O \xrightarrow[\text{acid}]{\text{dilute}} CH_3\text{-}\overset{OH}{CH}\text{-}\overset{OH}{CH_2}$$

(Glycol is from the Greek route, glyk-, meaning sweet. The link is through the sugars, which have structures much like propylene glycol, with multiple carbons and hydroxyl radicals.)

Material Balance	
Feed:	
Propylene oxide	887 lbs.
Water	275 lbs.
Catalyst	trace
Product:	
Propylene glycol	1000 lbs.
Dipropylene glycol	130 lbs.
Tri-, tetra-, and heavier glycols	32 lbs.

The hardware for propylene glycol is the same as that shown in Figure 10–5. Just substitute propylene for ethylene to identify the streams.

Commercial Aspects

Uses. Polyester resins use up about 60% of the propylene glycol (and most of the dipropylene glycol) manufactured. The remainder is used as a tobacco and cosmetic humectant (a chemical that keeps moisture around), automotive antifreeze and brake fluid ingredients, food additive, a plasticizer for various resins, and making coatings.

Many of the derivatives of propylene glycol, namely the ethers and the acetates, behave very much like the corresponding ethylene glycol derivatives. For that reason they are easily substitutable for each other.

Propylene Glycol Properties	
Freezing point	76.0°F (–60.0°C)
Boiling point	361.1°F (183.7°C)
Specific gravity	1.0381 (heavier than water)
Weight per gallon	8.72 lbs./gal.

Properties and Handling. You can tell from the applications that propylene glycol is safe. It is non-toxic, non-flammable, and even fit for human consumption (in small doses). It is a colorless, odorless, sweet tasting liquid, completely miscible or soluble in water. Propylene glycol is available in three grades: NF (99.99%), technical (99%), and industrial (95%).

Chapter XI in a nutshell...

Propylene oxide is a triangle-shaped cyclic compound with a $-CH_3$ group "along the bottom." An oxygen atom bonded to two adjacent carbon atoms makes the triangle. No catalyst has been found to efficiently oxidize propylene directly, so the commercial processes involve two steps. The chlorohydrin route uses propylene, chlorine, and oxygen. The indirect oxidation route involves isobutane, air, and propylene. The latter also produces tertiary butyl alcohol, a valuable gasoline blending component.

Propylene oxide is a liquid at room temperature. It readily converts to propylene glycol in the presence of water and a little acid. But the majority of propylene oxide is used to make polymers, including polyurethane foam.

Propylene glycol looks like propane with hydroxyl groups, -OH, substituted for a hydrogen on two adjacent carbons. Propylene glycol also is used to make polymers, and in a variety of smaller applications such as solvents, humectants, and food additives.

Exercises

1. Fill in the blanks:
   a. Forming a ringed compound like propylene oxide from propylene is called ___________.
   b. Two commercial routes to propylene oxide are ___________ and ___________ ___________. The ___________ ___________ route has not been commercially successful.
   c. Commercially valuable by-products of propylene oxide and of propylene glycol manufacture mentioned in this chapter are ___________, ___________, and ___________.
2. Why is the chlorohydrin route losing favor as the preferred route to propylene oxide?
3. Write the equation for the indirect oxidation of ethylbenzene to propylene oxide and styrene.

COMMENTARY TWO

"Is you deaf?"
sez Brer Rabbit sezee.
"Kaze if you is,
I can holler louder,"
sezee.

■

The Wonderful Tar-Baby Story
Joel Chandler Harris (1848–1908)

REVIEW

The "Mutt and Jeff" chemicals introduced in the last five chapters can be considered first and second derivatives of the petrochemical building blocks ethylene, propylene, and benzene, as shown in the following figure. The first line derivatives are all catalyst induced. Pressure and temperature are the operatives. Styrene and vinyl chloride are a little special. The characteristically reactive double bond in both of them is induced by high temperature pyrolysis, cracking off the required hydrogen atoms. That makes these second level derivatives behave much like the basic building blocks ethylene and propylene. In fact their major applications are in polymers, similar to much of the two olefins' volume.

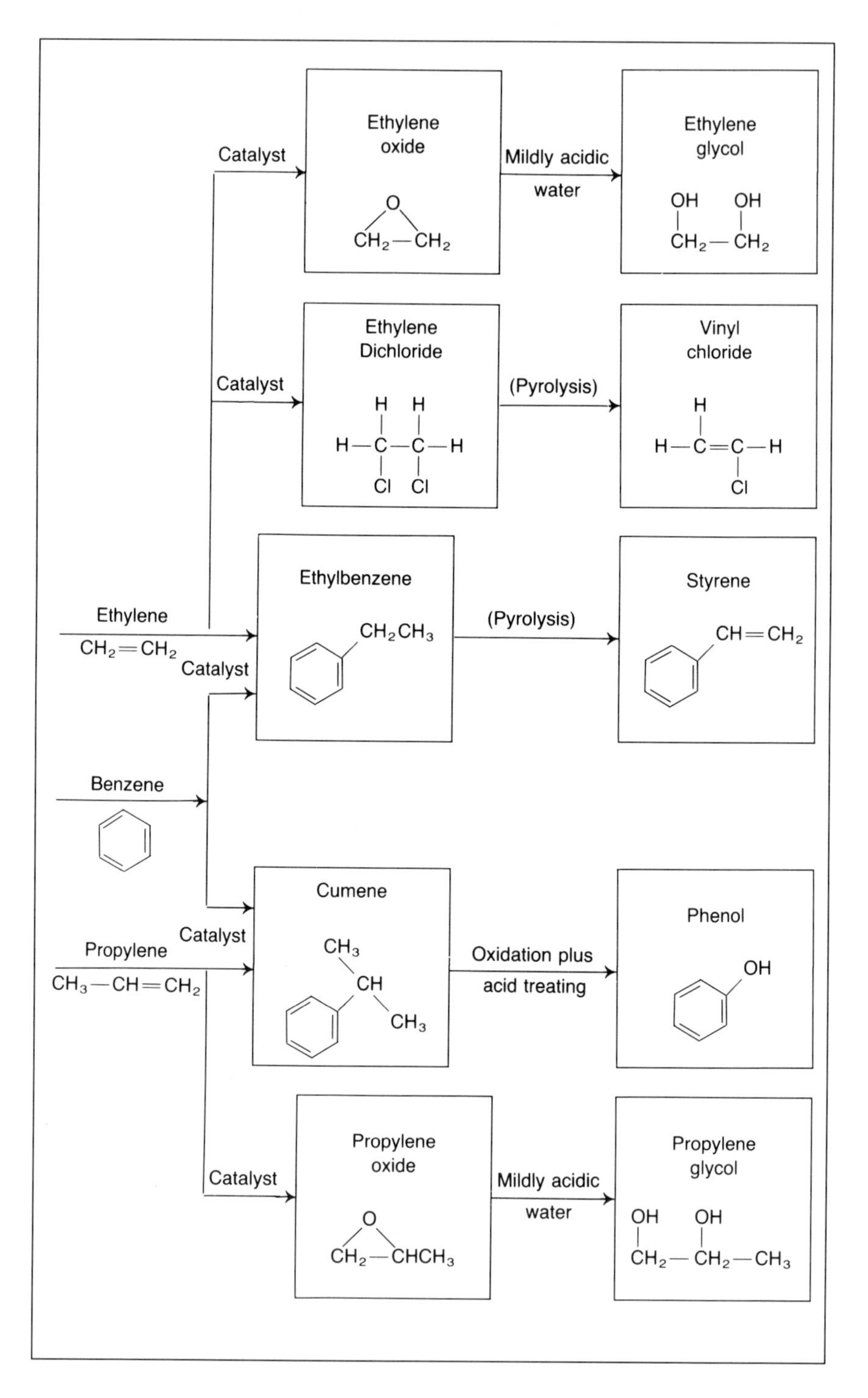
Ethylene
$CH_2{=}CH_2$
Catalyst
Ethylene oxide
$CH_2{-}CH_2$ (O bridge)
Mildly acidic water
Ethylene glycol
OH OH
$CH_2{-}CH_2$
Catalyst
Ethylene Dichloride
H H
H—C—C—H
Cl Cl
(Pyrolysis)
Vinyl chloride
H
H—C=C—H
Cl
Catalyst
Ethylbenzene
CH_2CH_3
(Pyrolysis)
Styrene
$CH{=}CH_2$
Benzene
Catalyst
Propylene
$CH_3{-}CH{=}CH_2$
Cumene
CH_3
CH
CH_3
Oxidation plus acid treating
Phenol
OH
Catalyst
Propylene oxide
O
$CH_2{-}CHCH_3$
Mildly acidic water
Propylene glycol
OH OH
$CH_2{-}CH_2{-}CH_3$

There is another route to propylene oxide and to styrene not shown on the figure. Indirect oxidation of iso-butane or of ethylbenzene just doesn't fit neatly on the chart, but the former is a commercial process yielding PO and TBA; the latter gives PO and styrene.

FOREWORD

The next six chapters cover a collection of petrochemicals not particularly related to each other. Synthesis gas is a basic building block. The alcohols in Chapters XII and XIII, the aldehydes in XIV, the ketones in XV, and the acids in XVI are all close to each other in the petrochemical family tree. Maleic anhydride, acrylonitrile, and the acrylates . . . ? Well, they're all used to make polymers, and they had to be somewhere.

METHANOL AND SYNTHESIS GAS

> As the poet said, "Only God can make a tree"—probably because it's so hard to get the bark on.
>
> ▪
>
> ***Without Feathers***
> **Woody Allen, 1935-**

The order in which you have to approach these two subjects is reversed in the chapter title because you might be apprehensive if they weren't. After all, everyone knows what methanol is. It's methyl alcohol, CH_3OH, wood alcohol, carbinol, or if you're a student of medieval culture, aqua vitae. But what is "synthesis gas?" It's not a familiar name because it's not usually handled in commercial transactions. The term synthesis gas refers to various mixtures of carbon monoxide (CO) and hydrogen (H_2) used for the manufacture of certain petrochemicals. In the nineteenth century, it was produced by passing steam over coke at very high temperatures. Today it's made largely from natural gas (methane). But a few paragraphs about synthesis gas, how it's made, and how it can be used in the synthesis of other petrochemicals will be beneficial. That's particularly true because two important chemicals, ammonia and methanol, are derived from synthesis gas.

SYNTHESIS GAS

Mother Nature hasn't provided any convenient sources of pure CO and H_2. There's some of each contained in natural gas, but usually not in sufficient quantities to justify going after it. But these two compounds, either in the combined state or separate, are readily convertible to a number of commercial compounds. With that as a motivator, several processes have been developed to convert natural gas to synthesis gas. Natural gas is largely methane (CH_4), and that provides a source of carbon and hydrogen. Air or water provide the other necessary ingredient, oxygen.

Synthesis gas can easily be confused with the oxymoron synthetic natural gas, SNG. Both are sometimes called "syngas." But SNG is basically methane made from petroleum products, like naphtha or propane, or from coal. It's used as a substitute for or supplement to natural gas.

The Synthesis Gas Processes

The two predominant methods of making synthesis gas are steam reforming and partial oxidation. Both are quite simple. The steam reforming method involves passing methane or naphtha plus steam over a nickel catalyst. The reaction, if methane is the feedstock, is:

$$CH_4 + H_2O \longrightarrow CO + 3H_2$$

The reaction relies on the brute force of high temperatures and pressures and must be carried out in hardware much like the cracking furnaces described in the ethylene chapter. As always with cracking, undesirable reactions occur, resulting in the formation of CO_2 and carbon. The latter is particularly a nuisance because it sets down on the catalyst and deactivates it.

The other method is the partial oxidation of methane:

$$CH_4 + \frac{1}{2} O_2 \longrightarrow CO + 2H_2$$

Like the steam reforming method, this process takes place at severe conditions, high temperatures and pressures, but no catalyst. The reaction is called partial oxidation because it is kept from going to CO_2 by limiting the amount of oxygen fed to the process.

The partial oxidation method is normally used for heavier feedstocks, everything from naphtha to residual fuel, in those places where natural gas or light hydrocarbons (ethane, propane, or butane) are not readily available.

The yield of CO is not 100% in either process. You can see in Table 12–1 that plenty of CO_2 also gets formed as a by-product.

Table 12–1 Synthesis Gas Composition (Percent yield based on methane feed)

	H_2	CO	CO_2	Total
Steam reforming	75	15	10	100
Partial oxidation	50	45	5	100

Fortunately, CO_2 can be removed without too much difficulty by solvent extraction. Even better, it can then be reacted with steam and more methane to give CO and H_2. This step is also done at high temperatures and pressures, and a nickel catalyst is used.

$$3CH_4 + CO_2 + 2H_20 \longrightarrow 4CO + 8H_2$$

In addition, this step is sometimes used to supplement the other reactions to get the proper combination of CO and H_2, since its CO:H_2 ratio is so different. Methanol production, as discussed below, requires this help.

Synthesis gas can be tailored in this manner to fit any number of specific applications. For example, a commercial route to aldehydes (the R-CHO signature group) and alcohols (the R-OH signature group) is the Oxo reaction, as discussed in Chapters XIII and XIV. In that reaction, the CO:H_2 ratio needed is 1:1. Careful adjustment of the three feedstocks, CH_4, CO_2, and H_20 and the amount of recycling gives the proper combination.

Commercial Aspects

Most of the synthesis gas produced is captive. That is, it's consumed by the manufacturer. Synthesis gas plants are normally integrated into the adjacent application plant. When there is a two-party transaction involved, the properties of the synthesis gas stream are normally specified in a contract. There are no universally accepted standards that apply with this stream.

The only practical way to move synthesis gas around is by pipeline, and even in two-party transactions the pipelines are usually no longer than a mile or two. Beyond that, the pipeline capital cost starts to affect the economics of the applications.

AMMONIA

The most important use of synthesis gas is the manufacture of ammonia (NH_3) via the Haber process. A mixture of nitrogen and hydrogen are passed over an iron catalyst (with aluminum oxide present as a "promoter"). The operating conditions are extreme—800°F and 4000 psi.

$$N_2 + 3H_2 \longrightarrow 2NH_3$$

Why synthesis gas? And where does the nitrogen come from? Synthesis gas, of course, provides the hydrogen; air provides the nitrogen. And if the synthesis gas process is partial oxidation, then there was probably an air separation plant associated with it. That separates the oxygen from the nitrogen for making the synthesis gas, and leaves the nitrogen for feed to the ammonia plant.

In most ammonia plants, there are facilities to remove CO from the feed because CO will poison the catalyst. Generally, the technique used is to react the CO with water to produce CO_2 and H_2. The CO_2 is removed by solvent extraction, and the H_2 is recycled. (In case you were wondering, typical solvents used to remove CO_2 are ethanolamine or an aqueous solution of potassium carbonate.)

METHANOL

There's a good reason why methanol is commonly called wood alcohol. The early commercial source was the destructive distillation of the fresh-cut lumber from hardwood trees. When wood is heated without access to air to temperatures above 500°F, it decomposes into charcoal and a volatile fraction. Among the compounds in the volatile fraction is methanol. Hence, the name wood alcohol or wood spirits.

Since 1923, methanol has been made commercially from synthesis gas, the route that provides most of the methanol today. The plants are often found adjacent to or integrated with ammonia plants for several reasons. The technologies and hardware are similar, and the methanol plant can use the CO_2 made in the Haber ammonia process. In this case the route to methanol is to react the CO_2 with methane and steam over a nickel catalyst to give additional CO and H_2 and then proceed to combine these to make methanol:

$$3CH_4 + 2H_2O + CO_2 \longrightarrow \underset{\text{synthesis gas}}{4CO + 8H_2}$$

$$CO + 2H_2 \rightleftharpoons \underset{\text{methanol}}{CH_3OH}$$

The double arrows in the methanol reaction indicate that the reaction can go in either direction. There is a principle here that is taught in the sophomore P-chem class (Physical chemistry) of every chemical engineer. Methanol in the vapor state takes up only one-third the volume as the equivalent amounts of CO and H_2. So in order to "push the reaction to the right," the process is run under pressure. That causes the compound that takes up less volume to be favored . . . synthesis gas to methanol rather than methanol to synthesis.

The first commercial plant that converted synthesis gas to methanol was built in 1924 in Germany by BASF. It ran at very high pressures (3500–5000 psi) and used a zinc-copper catalyst. In the years since then, further development of catalysts has brought the pressures down, eliminating much of their expensive capital and operating costs. In the 1950s, medium pressures of 1500–3500 psi were in vogue. At the present, newer catalysts based on copper-zinc oxide have resulted in lower pressures of 500–1500 psi in 90% of the plants.

The process is still expensive, and that continues to give incentives to ongoing research to find the elusive catalyst that will permit direct conversion of methane to methanol without having to break apart the methane and reassemble it again. Breakthoughs in this technology are possible at any time which could obsolete this whole sector of the petrochemical industry. While that's true of many parts of the industry, this process seems somewhat more vulnerable.

THE PLANT

The process for synthesis of methanol involves these basic steps:

1. Steam reforming of natural gas plus addition of CO_2 to adjust the $CO:H_2$ ratio to 2:1
2. Compression to 500 to 1500 psi
3. Synthesis in a catalytic converter
4. Purification—distillation.

The hardware is shown in Figure 12–1. To protect the compressors, a water knock-out column in front is necessary. It keeps water slugs from forming during compression, sending turbine blades flying all around the plant.

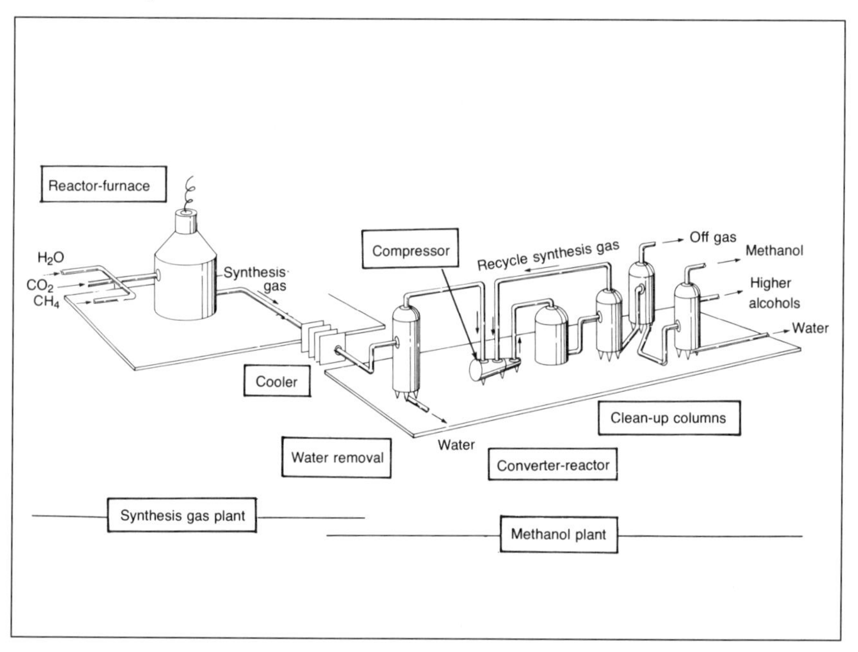

Fig. 12–1 Synthesis gas and methanol plants

The compressed gas is heated and passed through a reactor that has baskets of catalyst. In between the baskets are heat exchangers. The reaction is exothermic, but the reaction is sensitive to the temperature, so heat must be rapidly removed.

The effluent from the reactor contains only 5–20% methanol because the one-pass conversion is very low. After cooling and pressure let-down, the liquid methanol can be removed and further purified by distillation. The unreacted synthesis gas is recycled to the reactor.

Methanol of 99% purity is obtained. By-products are 1–2% dimethyl ether (CH_3OCH_3), about 0.5% higher alcohols (ethyl, propyl, isobutyl, and higher), and some water.

Material Balance

Feed:	
Carbon monoxide	921 lbs.
Hydrogen	132 lbs.
Product:	
Methanol	1000 lbs.
By-products	53 lbs.

Commercial Aspects

Properties and Handling. Methanol is a colorless, volatile liquid at room temperature with an alcoholic smell. It mixes with water in all proportions and burns with a pale blue flame. Methanol is highly toxic. As little as a fifth of a shot (10cc) can cause blindness. Larger amounts kill. It should never be applied to the body as a rubbing alcohol because the vapors are so toxic.

Sales grades of methanol include 95 and 97% purity. U.S. Federal grade must meet 99.8% minimum purity and be acetone free. Methanol can be transported in conventional tank trucks, tank cars, ships, and barges, but in closed systems. The red hazardous materials markings are required.

Methanol Properties

Freezing point	–143.7°F (–97.6°C)
Boiling point	148.3°F (64.6°C)
Specific gravity	0.792 (lighter than water)
Weight per gallon	6.59 lbs./gal.

Uses. About 40–50% of the methanol made is converted to formaldhyde. That's not because the embalming business is so good. Formaldehyde is a feedstock for amino and phenolic resins, which are used as adhesives in plywood and in the automotive and appliance industry to make parts (all the agitators in washing machines used to be made out of phenolic resins); for hexamethylene tetramine, used in electronic plastics; pentaerythritol, used for making enamel coatings and for floor polish and inks; for butanediol, a chemical intermediate; and acetic acid, which is widely used itself as a feedstock and solvent and warrants its own treatment later on. In the textile business, formaldehyde is used to make fire retardants, mildew resistant linens, and permanent press clothing.

Another application of methanol is methyl chloride, used in making silicone rubber, including the caulking and sealing compounds that will set at room temperatures (the kind you buy in a tube at the hardware store).

The fastest growing use of methanol is not in the petrochemical industry at all, but in the automotive fuel business. Methanol gets in via two routes, through the manufacture of MTBE (methyl tertiary butyl ether), a gasoline blending compound, and as a direct substitute for gasoline—either in part as a gasohol blend or in total.

Atmospheric pollution problems have focused public policy on the elimination of the pollutants formed during the combustion of gasoline in highway vehicles. The formation of ozone, nitrogen oxides, and some sulfur oxides is thought to be alleviated by the addition of organic compounds that contain oxygen in them, such as MTBE and methanol. Commercial testing of gasoline blends with these compounds and local laws requiring minimum content became prevalent in the U.S in the late 1980s.

As motor fuels, MTBE and methanol have different appeals. MTBE has a high octane number, which makes it suitable for blending premium gasoline. Methanol has a lower octane number but is cheaper to manufacture. Still it is generally considered more expensive than gasoline on an economic basis as long as the price of crude is under about $30.00 per barrel. A gallon of methanol, by the way, is not equivalent to a gallon of gasoline. The energy content of gasoline is about 1.8 times that of methanol, and the miles per gallon in an equivalent engine are in about the same proportion.

Methanol is also used to "de-nature" ethanol. There's not much difference between synthetic ethyl alcohol and the "real thing" made from rye and other grain. Methanol is added, for political reasons, up to 10%, to the synthetic stuff to keep it from being substituted for the "real" or "natural" ethyl alcohol. Ethanol "de-natured" in this way is toxic enough to cause headaches, dizziness, vomiting, blindness, and coma, depending on how much is consumed. That's usually a sufficient threat.

Smaller volumes of methanol are used in the production of dimethyl terephthalate which goes into polyester fibers; in methyl methacrylate, which goes into plastics.

Chapter XII in a nutshell...

Synthesis gas is a loose name for hydrogen/carbon monoxide mixtures of varying proportions. These two compounds are so basic, they are too simple to start with for most petrochemicals. The primary applications of synthesis gas are only ammonia and methanol manufacture, and a little normal butyl and 2-ethylhexyl alcohol production.

Synthesis gas is made by decomposing methane (from natural gas) in the presence of water. The reaction takes place at high pressure and temperature in the presence of a catalyst. The proportion of H_2 and CO depends on the amount of CO_2 that is left in the product stream or is recycled to be converted to CO/H_2. Most synthesis gas plants are built adjacent to the plants where the synthesis gas will be used.

Methanol, CH_3OH, the simplest alcohol, is made by reacting CO and H_2 at high pressure over a catalyst. Methanol is a liquid at room temperature and is highly toxic. It is used to make formaldehyde, acetic acid, and other chemical intermediates. It is used as a feedstock for MTBE (methyl tertiary butyl ether), a valuable gasoline blending component.

Exercises

1. You need two H_2 molecules for each CO molecule to make methanol, CH_3OH. How do you get CO and H_2 in nearly the right proportions so you waste as little as possible of either?
2. What's the synergy or affinity of ammonia and methanol plants?
3. What are the largest applications of methanol?

THE OTHER ALCOHOLS

" 'Twas a woman that drove me to
drink, but I never had the
courtesy to go back and thank her."

■

W. C. Fields, 1879–1946

There are many other commercial alcohols besides methanol. This chapter treats the ones traded in the largest volumes, ethyl alcohol, isopropyl alcohol (IPA), normal butyl alcohol (NBA), and 2-ethyl hexanol (2-EH). You'll also get a brief discussion of the plasticizer and detergent range alcohols, the C_8 to C_{18} alcohols.

A good way to think about alcohols is to start with water, H_2O, which can be written H-OH. An alcohol is formed if the H is replaced by an organic grouping. The chemist's way of referring to any organic grouping like a chain or a ring is the symbol R. So, the alcohol signature is R-OH.

Often, but not always, the alcohol is named

after whatever R is. CH_3-CH_2-OH is ethyl alcohol, CH_2=CH-OH is vinyl alcohol, but C_6H_5-OH, a benzene ring with a hydroxyl group, is phenol. (See Fig. 13–1.)

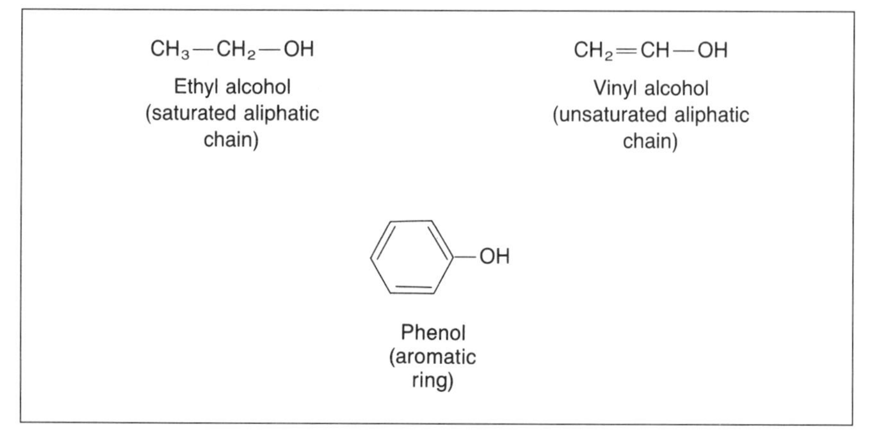

Fig. 13–1 Alcohols

By the way, when there are more than one hydroxyl group per molecule, it is still an alcohol. Ethylene glycol

$$\underset{\text{OH}}{\text{CH}_2}\text{-}\underset{\text{OH}}{\text{CH}_2}$$

is a polyhydric alcohol, as is glycerine

$$\underset{\text{OH}}{\text{CH}_2}\text{- }\underset{\text{OH}}{\text{CH}}\text{- }\underset{\text{OH}}{\text{CH}_2}$$

or glucose

$$\underset{\text{OH}}{\text{CH}_2}\text{- }\underset{\text{OH}}{\text{CH}}\text{- }\underset{\text{OH}}{\text{CH}}\text{- }\underset{\text{OH}}{\text{CH}}\text{- }\underset{\text{OH}}{\text{CH}}\text{-CHO.}$$

Like Caesar's Gaul, petrochemical processes used for manufacturing alcohols today *est divida in tres partes.*

1. Hydration. The addition of water to an olefin:

   $$\underset{\text{ethylene}}{CH_2=CH_2} + H_2O \longrightarrow \underset{\text{ethyl alcohol}}{CH_3\text{-}CH_2\text{-}OH}$$

2. Oxo reaction. Reacting an olefin with synthesis gas (CO and H_2),

followed by hydrogenation (addition of hydrogen), producing an alcohol containing one more carbon than the original olefin.

$$\text{Olefin + Syngas} \longrightarrow \text{R-CHO}$$

then,

$$\text{R-CHO} + H_2 \longrightarrow \text{R-}CH_2\text{-OH}$$

3. Ziegler reaction. Produces an even number, straight chain alcohol in carbon number range C_8 to C_{18}. The reaction involves "growing" chains of ethylene on an aluminum-organic compound, adding the -OH using water, then "clipping off" the alcohol.

$$\underset{\text{Triethyl aluminum}}{Al(C_2H_5)_3} + \underset{\text{ethylene}}{CH_2{=}CH_2} \longrightarrow \underset{\text{a polymer chain}}{\text{Trialkyl aluminum}}$$

then,

$$\text{Trialkyl aluminum} + O_2 \xrightarrow{H_2O} \underset{\text{alcohol}}{\text{R-OH}} + \underset{\text{aluminum hydroxide}}{Al(OH)_3}$$

where R is a C_8 up to C_{18} alkyl group.

The best way to elaborate on these three processes is to look at specific alcohols.

ETHYL ALCOHOL

The fermentation of sugar in the presence of yeast to produce ethyl alcohol in the form of wine goes back beyond historians' recorded words. The sugar came from grapes. Later, starch from grain, potatoes, or "corn squeezins" was used also. The yeast came from living matter in the form of mold or fungus. Yeast contains the enzyme "zymase." It's this enzyme that catalyzes the fermentation of sugar. Mix sugar (in grape juice) with yeast and they will react slowly—weeks, months, maybe years, to form ethyl alcohol and carbon dioxide, as well as minor amounts of some aldehydes. Depending on preferences, some of the non-alcoholic contents can be separated by distilling.

Alcoholic beverages in the U.S. are made exclusively by the fermentation process, not the petrochemical process. It has nothing to do with the chemistry. It's a law enacted to protect the grain growers, not the consumers.

The convention for identifying the alcoholic content of beverages is "proof." One hundred proof is 50% ethyl alcohol; 86 proof scotch is 43% ethyl alcohol, and so on. Divide by two. So pure ethyl alcohol is 200 proof.

Until World War I, fermentation accounted for all the ethyl alcohol produced in the U.S. In 1919, a petrochemical route based on ethylene, sulfuric acid, and water was developed commercially and called indirect hydration. By 1935, only 10% of the ethyl alcohol was produced this way, primarily because of the expense of the ethylene at that early stage of the industry. With the rapid improvements of ethylene technology, the share quickly grew to 90% by the 1960s. At that time an alternate route, direct hydration, was developed eliminating the use of sulfuric acid and one step in the process. Direct hydration virtually replaced all the indirect hydration process by the 1970s. Advantages were higher yields, less pollution, and lower plant maintenance due to less corrosion—all leading to better economics. Currently, about 95% of the domestic ethyl alcohol is produced via the direct catalytic hydration of ethylene.

The Process

The chemical reaction,

$$CH_2{=}CH_2 + H_2O \longrightarrow CH_3\text{-}CH_2\text{-}OH$$

takes place in a single reactor, as shown in Figure 13–2. The rest of the facilities are handling and cleanup hardware.

Ethylene is compressed to 1000 psi, mixed with water, and heated to 600°F. The two reactants, both in a vapor phase, are fed down a reactor filled with catalyst. The catalyst is phosphoric acid on a porous inert support (usually diatomaceous earth or silica gel).

The ethylene conversion to ethyl alcohol per pass through the reactor is only 4 to 6%, so most of the ethylene needs to be recycled. But first the reactor effluent is cooled and caustic washed. As the effluent cools down, the ethyl alcohol liquefies, and the ethylene can easily be separated. It is then "scrubbed" by sloshing it through water prior to recycle.

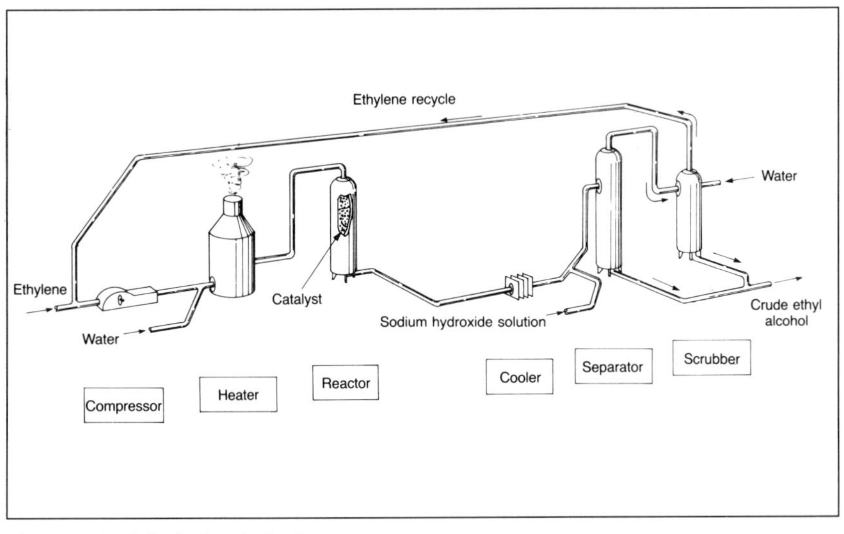

Fig. 13–2 Ethyl alcohol plant

The mixture from the bottom of the separator and the scrubber is crude ethyl alcohol. That is, it contains the ethyl alcohol, water, and all the by-products. Further distillation separates out an ethyl alcohol-water mixture (95% ethyl alcohol, 5% water) that boils at a single, constant temperature, called an azeotrope. Now that presents a special, knotty problem. Since the mixture boils at a temperature lower than ethyl alcohol, how does the ethyl alcohol get separated from the water? Not by ordinary distillation.

The answer is like fighting fire with fire—another azeotrope is formed. When benzene is added to ethyl alcohol and water, a ternary azeotrope, a mixture of three compounds that boil at a single temperature, is formed. The ternary azeotrope has the composition of 68% benzene, 24% ethyl alcohol, and 6% water, and it boils a temperature lower than the binary ethyl alcohol/water azeotrope. So when a little benzene is added to the ethyl alcohol/water mixture and then put through a distillation column, the ternary azeotrope, in the composition just mentioned, will come off the top, taking with it all the benzene, all the water, but just some of the ethyl alcohol. Out the bottom comes what's left, the rest of the ethyl alcohol in nearly pure form. Slick.

The ternary azeotrope is liquified and processed further to recover and recycle the benzene and the ethyl alcohol content and to reject the water.

By-products often mixed with the ethyl alcohol are diethyl ether (an anesthetic) and acetaldehyde, both of which can be easily hydrogenated to ethyl alcohol.

Material Balance

Feed:	
Ethylene	640 lbs.
Water	412 lbs.
Catalyst	small
Product:	
Ethyl alcohol	1000 lbs.
By-products	52 lbs.

Commercial Aspects

Uses. Ethyl alcohol has been used as a chemical intermediate in the manufacture of acetaldehyde and acetic acid. This application is fading fast because of recent process improvements in direct oxidation of ethylene. (Over 80% of today's acetaldehyde is made via the direct oxidation route.)

Ethyl alcohol is still in demand as a solvent in personal care products (after-shave lotion, mouthwash), cosmetics, pharmaceuticals, and even surface coatings. It is a precursor for ethyl chloride, which is used to make tetraethyl lead, diethyl ether (the anesthetic), and ethyl acetate, a solvent for coatings and plastics. (When you smell nail polish remover, you're smelling ethyl acetate.)

Ethyl alcohol is being used extensively in the U.S. as an automotive gasoline supplement, known as gasohol. However, the source of the ethyl alcohol is not the petrochemical industry but the fermentation industry. Special tax incentives have been given to manufacturers of ethyl alcohol made from grains or corn, and that has made the process competitive with oil base gasoline. Those incentives are not available to petrochemical sources of ethyl alcohol, and the latter route will remain non-competitive. Even if oil prices increase, ethyl alcohol feedstock costs are likely to increase simultaneously and in proportion, leaving the gasoline market economically unreachable by ethyl alcohol from petrochemical sources.

Properties and Handling. Ethyl alcohol is a colorless, flammable liquid (good for flambe') having a characteristic odor nearly universally recognizable. It is soluble in water (and club soda) in all proportions. It's commercially available as 190 proof (the 95% ethyl alcohol-water azeotrope) and "absolute" (200 proof). It is frequently denatured to avoid the high tax associated with 190 and 200 proof grades. Methanol and/or sometimes formaldehyde are common denaturants used to prevent consumption as an alcoholic beverage.

Because of the flammability, ethyl alcohol is transported as a hazardous material.

Ethyl Alcohol Properties	
Boiling point	179.2°F (78.3°C)
Freeze point	−137.9°F (−114.1°C)
Specific gravity	0.789 (lighter than water)
Weight per gallon	6.58 lbs./gal.

ISOPROPYL ALCOHOL

Technically, iso-propyl alcohol (IPA) can be made via the direct hydration method, but the severe operating conditions required make the route energy-intensive and "therein lies the rub." The preferred route to IPA is indirect hydration. It is more economical than the direct hydration route because lower pressures and temperatures are involved. (Ethylene won't react at the lower temperatures, so this route isn't practical for ethyl alcohol.)

The specifications for the feed to the IPA plant can be loose. Refinery grade propylene, even with some small amounts of ethane and ethylene can be used, because the C_2's and propane don't react. They just pass through the process. As a matter of fact, the process acts as kind of a C_3 splitter, since about 50% of the propylene gets converted to IPA in each pass through the reactor, leaving high purity propane behind.

The Process

Propylene is absorbed by concentrated sulfuric acid to form isopropyl hydrogen sulfate. That's subsequently hydrolyzed with water to IPA and dilute sulfuric acid.

The propylene stream is fed into the bottom of a column (Fig. 13–3) packed with baffles to give intimate contact. Concentrated (85%) sulfuric acid is introduced at the vessel top.

As the acid and propylene slosh past each other, about 50% of the propylene reacts with the sulfuric acid to form the sulfate. The reaction is exothermic, so the contents of the tower must be continually cooled to maintain a 70–80°F temperature. This minimizes by-products, particularly propylene polymers. Any higher olefins, usually C_4 and C_5, in the propylene feed will be absorbed by the sulfuric acid, forming sulfates and bi-sulfates. They have to be removed in the clean up facilities. The yield of IPA from propylene, that is the proportion of propylene that ends up as IPA, is about 70%.

The propylene from the reactor top can be recycled to the feed (though it's not shown that way in Figure 13–3.) The concentration might have to be boosted by a splitter or by the addition of some chemical grade propylene. The effluent from the reactor bottom is dumped into a lead-lined tank and diluted with water and steam, cutting the unreacted sulfuric acid to about 20%. Mixing sulfuric acid and water is exothermic, and that heat plus a little steam is sufficient to hydrolyze the isopropyl hydrogen sulfate to IPA. With a little more steam, the crude IPA flashes (vaporizes) out of the dilution tank and goes to a fractionator for concentration. The dilute H_2SO_4 stream is sent off to be cleaned up and reconstituted to higher concentration for re-use.

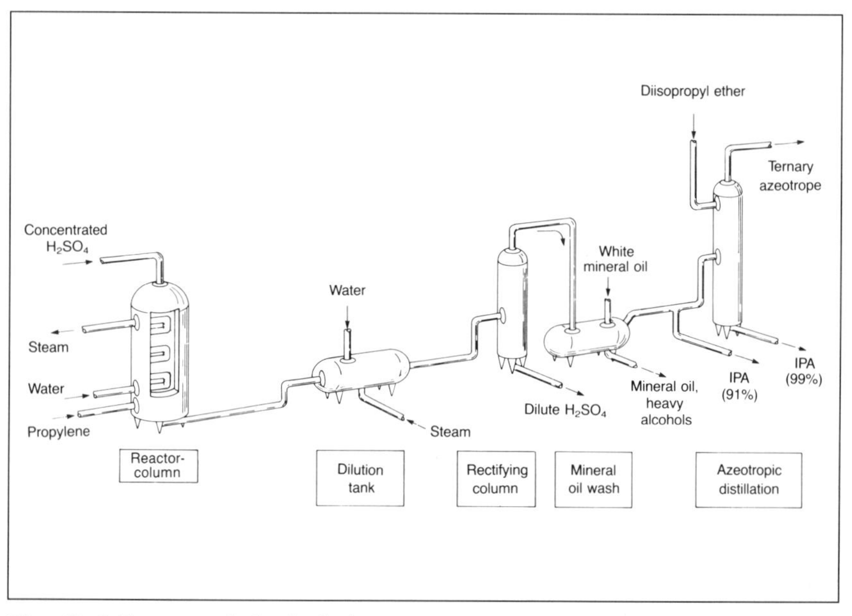

Fig. 13–3 Isopropyl alcohol plant

At the fractionator, a 91% IPA/water azeotrope distills from the top, carrying along most of the other organics. The IPA/water azeotrope is washed with mineral oil to remove the C_4's, C_5's and higher alcohols. It's further treated with sodium hypochlorite to give the water-white technical grade, which is still the 91% IPA/water azeotrope.

Like ethyl alcohol, the absolute (99 + %) grade is made by forming a ternary azeotrope. In this case, DIPE (di-isopropyl ether) is used to form the ternary with water and IPA. But the idea is exactly the same.

Material Balance	
Feed:	
Propylene	900 lbs.
Water	385 lbs.
Sulfuric acid (85%)	1235 lbs.
Product:	
Isopropyl alcohol	1000 lbs.
Sulfuric acid	1235 lbs.
By-products	285 lbs.

You may wonder why the process produces isopropyl alcohol instead of (normal) propyl alcohol. With the exception of ethylene, direct or indirect hydration of an aliphatic olefin always produces an alcohol with the hydroxyl group preferentially attached to the double-bonded carbon with the least number of hydrogen atoms.

Commercial Aspects

Uses. In 1980, over 50% of the IPA produced was used to make acetone (dimethyl ketone). By the end of the 1980s, the percent was down to less than 10. The plants that co-produce phenol and acetone had almost entirely replaced the IPA-to-acetone route.

Now IPA is used primarily as a coating and processing solvent in paints, electronics applications, synthetic resins, personal care products, and cosmetics. It also is used as a chemical intermediate for isopropyl esters and hydrogen peroxide production.

At one time IPA was used as a gasoline additive to prevent cold weather stalling, but it has been largely displaced by DIPE.

And of course, IPA is used as rubbing alcohol, because of its innocuous non-toxic odor, its low boiling (vaporization) temperature, and moderate heat of vaporization. It "dries" rapidly, but won't give you frost bite like liquid butane might.

Properties and Handling. IPA is a colorless, flammable liquid with that characteristic, rubbing alcohol odor. It's soluble in water in all proportions, as well as most organic solvents. It is commercially available in technical grade (91%), chemical (98%), and absolute (99+%). Shipments by rail, truck, drum, etc., are routine, except that the flammability requires hazardous materials warnings.

Isopropyl Alcohol Properties	
Boiling point	180.5°F (82.5°C)
Freeze point	−129.1°F (−89.5°C)
Specific gravity	0.785 (lighter than water)
Weight per gallon	6.55 lbs./gal.

NORMAL BUTYL ALCOHOL AND 2-ETHYL HEXANOL

There's another convention that has been useful in describing alcohols, and that has to do with the positioning of the -OH or hydroxyl group. There are primary, secondary, and tertiary alcohols, depending on whether the hydoxyl group is attached to the primary, secondary, or tertiary carbon atom. In the case of the C_4 alcohols, the hydroxyl group can be connected to either:

- a primary carbon atom, one which is attached to only one other carbon atom;
- a secondary carbon atom, one which is attached to two other carbon atoms;
- a tertiary carbon atom, one which is attached to three other carbon atoms.

In other words, for the C_4 alcohols you can have:

CH_3-CH_2-CH_2-CH_2-OH — Normal butyl alcohol, NBA (1° alcohol, or R-CH_2-OH)

$$\begin{array}{l} CH_3\text{-}CH\text{-}CH_2\text{-}CH_3 \\ \quad\quad | \\ \quad\quad OH \end{array}$$

Secondary butyl alcohol, SBA (2° alcohol, or $\begin{array}{l} R\text{-}CH\text{-}OH \\ \quad\; | \\ \quad\; R \end{array}$)

$$\begin{array}{c} CH_3 \\ | \\ CH_3\text{-}C\text{-}0H \\ | \\ CH_3 \end{array}$$

Tertiary butyl alcohol, TBA (3° alcohol, or $\begin{array}{c} R \\ | \\ R\text{-}C\text{-}OH \\ | \\ R \end{array}$)

Normal butyl alcohol (NBA) was first recovered in the 1920s as a by-product of acetone manufacture via cornstarch fermentation. That route is almost extinct now. A small percent is still made from acetaldehyde. The primary source of NBA, however, is the Oxo process.

The Oxo process is used in a number of applications for extending the length of an olefin chain by one carbon. The reaction is between an olefin and synthesis gas (carbon monoxide and hydrogen, covered just in time in the last chapter) in the presence of a cobalt catalyst. It produces a mixture of aldehydes (the -CHO signature group), which readily undergo hydrogenation to alcohols. One important feature of the process is that it produces only primary alcohols. Most other petrochemical processes yielding alcohols produce secondary alcohols.

CH_3-CH=CH_2 + H_2 + CO ⟶ Normal butyraldehyde + Isobutyraldehyde
(Propylene, Synthesis Gas)

↓ add H_2

Normal butyl alcohol + Isobutyl alcohol

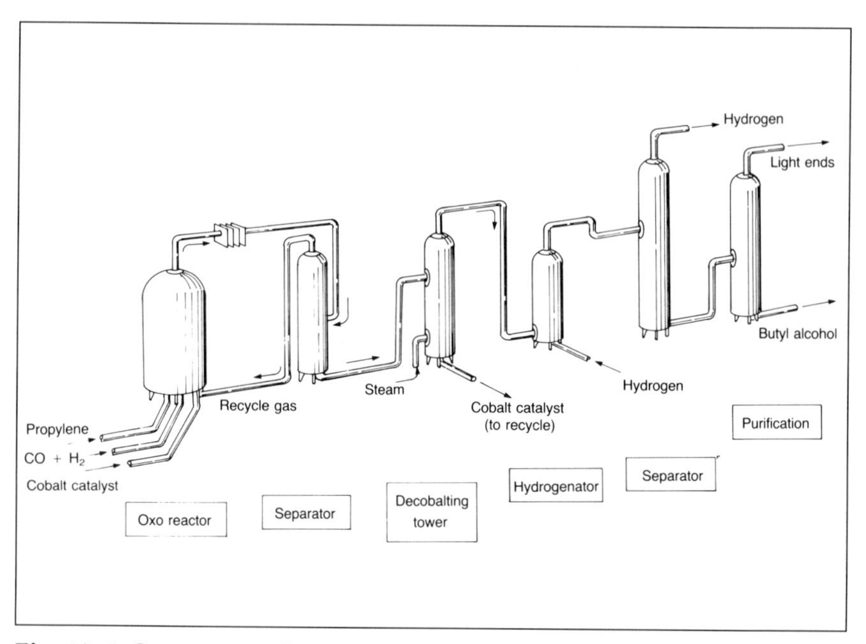

Fig. 13–4 Oxo process for normal butyl alcohol

Both of the butyl aldehydes and alcohols are formed in this process, but the demand for NBA greatly exceeds the isomer. By using mild operating conditions in the plant, the ratio of normal to iso- increases, though the total yield goes down. Recycling the iso-butyraldehyde to the feed "fools" the reaction also, and keeps down the iso- yield. Most processes yield an overall ratio of normal to iso- of about 4 to 1.

2-Ethyl hexanol, sometimes called 2-ethyl hexyl alcohol or more simply 2-EH, is one of the oldest high molecular weight aliphatic alcohols. What does it have in common with NBA? Both are made from propylene via the Oxo process, and both have the same aldehyde intermediate—normal butyraldehyde.

In the case of 2-EH production, the aldehyde dimerizes or reacts with itself. (Dimerize, remember, has the same Latin root -meros, as isomer, monomer, and polymer, and means part. A dimer is a chemical union of two molecules of the same compound.) The resulting C_8 dimer is also an aldehyde that can be hydrogenated to give 2-EH.

Dimerization:

normal butyraldehyde ⟶ "dimer" aldehyde

↓ add H_2

$$CH_3\text{-}CH_2\text{-}CH_2\text{-}\underset{\displaystyle CH_2\text{-}CH_3}{\underset{|}{CH}}\text{-}CH_2\text{-}OH$$

2-Ethyl Hexanol

The name 2-EH becomes apparent from the layout of the molecule. The straight C_6 chain gives the hexyl; the ethyl group, $-CH_2-CH_3$, gives the ethyl; and the ethyl group is connected to the second carbon from the right end of the chain, giving the two.

The Process for NBA

The feeds to an NBA plant are chemical grade propylene, synthesis gas, and an oil soluble cobalt salt (dissolved in the propylene), cobalt naphthenate. They are all fed to a reactor maintained at 250–300 °F and 3500–4000 psi. The reactor effluent contains unreacted gases, catalyst, and the aldehydes. The pressure is let down in a separator, and the unreacted gases are recycled. The hydrogenation step takes place in the conventional way, in a vessel packed with catalyst, where the aldehydes and hydrogen are admixed at 200–300 °F and 600–1200 psi. The catalyst is usually nickel or copper chromite on an inert carrier such as kieselguhr, silica gel, or alumina.

The crude butyl alcohols are finally separated and purified by distillation.

The Process for 2-EH

The same front-end process is used to make the butyraldehyde feed for 2-EH manufacture. But then the butyraldehyde is dimerized (sometimes called Aldol condensation when done in the presence of an alkaline catalyst). Some plants even combine the Oxo process and the Aldol process (then referred to as Aldox process).

In the Aldol route, the butyraldehydes and hydrogen are fed to a reactor filled with nickel catalyst, maintained at 300 °F and 2500 psi. Distillation of the effluent gives 2-EH yields of 93%.

It used to be that the iso-butyraldehyde had to be split out before the dimerization reaction, but catalysts improvements have permitted cogeneration of both C_4 alcohols and 2-EH.

Commercial Aspects

Uses. The motivation for first recovering NBA in the 1920s was its use as a lacquer solvent. That application is even stronger today. The NBA vapors from lacquer drying are non-toxic and virtually nonflammable. Other fast growing uses for NBA are plasticizers and chemical intermediates, mostly for esters and ethers used in water based coatings and adhesives systems.

The most important use for 2-EH is in the manufacture of di-2-ethyl hexyl phthalate (also known as dioctyl phthlate)—neither name would you want to try late on a Saturday night—which is used as a plasticizer to make polyvinyl chloride flexible. About 65% of the 2-EH goes to the manufacture of plasticizer. The growth of 2-EH acrylate is becoming an important application for 2-EH. This acrylate is finding use as a construction adhesive and in surface coatings. Other uses include industrial solvents, dispersing and wetting agents, and chemical intermediates.

Properties and Handling. NBA and 2-EH are non-volatile, colorless, non-toxic liquids, with relatively high boiling points. NBA is only slightly soluble in water and 2-EH is insoluble. A little rule of thumb here on solubility: "Like dissolves like." Methanol is very much like water, because the hydroxyl group in both (CH_3-OH and H-OH) is a significant part of the molecule. Same is true of ethyl alcohol and IPA. But as you get up to NBA, and especially 2-EH, the hydroxyl is minor, the carbon chain significant. The analogy between compounds and other solvents, including organics usually holds true—"Like dissolves like."

Both NBA and 2-EH are available in technical grade (98 to 99%) and are transported in normal equipment. No hazardous material label is required.

Normal Butyl Alcohol Properties	
Freeze point	−129.1°F (−89.5°C)
Boiling point	243.1°F (184.3°C)
Specific gravity	0.811 (lighter than water)
Weight per gallon	6.75 lbs. per gallon

2-Ethyl Hexanol Properties	
Freeze point	−94.0°F (−70.0°C)
Boiling point	363.7°F (184.3°C)
Specific gravity	0.834 (lighter than water)
Weight per gallon	6.94 lbs per gallon

SECONDARY AND TERTIARY BUTYL ALCOHOLS

Secondary and tertiary butyl alcohols (SBA and TBA) are produced by the absorption of butylene and iso-butylene in concentrated sulfuric acid. The processes are similar to the indirect hydrolysis route for IPA.

SBA is a colorless, high boiling point (212°F) liquid with a pleasant odor. TBA, on the other hand, is a white solid (melting point is 78°F) with a camphor-like odor. Both alcohols are traded as technical grade (99% purity) and need a hazardous (corrosive) materials label.

HIGHER ALCOHOLS

Alcohols in the carbon range of C8 to C18 are largely converted to derivatives and used primarily as plasticizers and detergents. As a result they are often referred to as plasticizer alcohols (the C8 to C12 alcohols) or detergent alcohols (the C12 to C18 alcohols).

There are three major routes to the higher alcohols—two purely synthetic and one that Mother Nature runs.

- Oxo process—the conversion of internal or alpha olefins. Alpha indicates the double bond is between carbon number one and two.
- Ziegler process—ethylene oligomerization.
- Hydrogenation of fatty acid esters derived from coconut oil, palm kernel oil, and tallow.

The Oxo process is the same as described a few paragraphs above—olefin plus synthesis gas combines to make an aldehyde then hydrogenated to give an alcohol. This process is used for the C_8 to C_{12} primarily, and seldom for the higher alcohols.

Ziegler alcohols, named after the brilliant chemist, Karl Ziegler, are even number carbon, straight chain, primary alcohols. That sounds like a very narrow description, and it is. There are no side chains coming out of this process because of the way the alcohols are "grown." The process involves the use of triethyl aluminum, $Al(C_2H_5)_3$, as a root (or a route) for adding on ethylene molecules. The growth of the molecule, or the polymerization, occurs so that ethyl groups just keep adding on to the end of other ethyl groups, with the result that the chains that are formed have 2, 4, 6 . . . carbon atoms in a straight chain.

$$\begin{array}{l} \quad CH_2\text{-}R_1 \\ \;/ \\ Al - CH_2\text{-}R_2 \\ \;\backslash \\ \quad CH_2\text{-}R_3 \end{array}$$

where R is the alkyl or ethyl group and the 1, 2, and 3 indicate different hydrocarbon chain lengths

The operating conditions and the catalyst characteristics determine the alkyl chain length, C_8 to C_{18} or even higher.

To get the alcohols, the hydrocarbon chains must be separated from the aluminum molecule. That's done by oxidizing the aluminum with air. Hydrolizing with water produces aluminum hydroxide (alum) and provides the hydroxyl to make the alcohol. The mixed or crude alcohols are washed and separated by distillation.

Natural-based alcohols are derived from the oils recovered from vegetable sources. These oils contain triglycerides, of the same kind that are fashionable for dieters to avoid these days. Because of that it is ironical that the chemical description of a triglyceride is glycerine esterified (made into an ester) with three moles of a fatty acid. Sounds like an underground plot to put weight on. In any event, the natural triglycerides are in abundance in those commodity oils, and with the help of methyl alcohol to separate the glycerine from the fatty acids, along with some mild hydrogenation, the alcohols are formed from the acids. Separation is again by distillation.

The term fatty acids, by the way, is a common term meaning thoses acids that are naturally found in animal or vegetable fats.

Uses. The higher alcohols are important raw materials for production of high molecular weight esters (phthalates, adipates, sulfates)

which are used as plasticizers and for conversion to anionic and non-ionic detergents. The big advantage of the straight hydrocarbon chains which are characteristic of these detergents is that they are biodegradable. As a result they are used extensively as light duty liquid dishwashing soaps, heavy duty powdered laundry soaps, and bar soaps.

Other minor uses are lacquer solvent, synthetic lubricants, perfumes, antifoaming agents, herbicides, and lube oil additives.

Chapter XIII in a nutshell...

Alcohols have the characteristic -OH signature group. There are a variety of ways to get that signature affixed to various hydrocarbons: synthesis gas is the source of oxygen and hydrogen as well as some of the other parts in methanol, normal butyl alcohol, and 2-ethyl hexanol; ethylene can be directly hydrated to ethanol (ethyl alcohol); and isopropyl alcohol from propylene requires a two step hydration.

Higher molecular weight alcohols are made from ethylene using the Oxo process—via ethylene oligomerization and hydration or direct hydrogenation of naturally occurring fatty acid esters in coconut and palm kernel oils and tallow. These alcohols are used in making plasticizers and biodegradable detergents.

Exercises

1. Why can't pure ethyl alcohol be obtained by simple fractionation of ethyl alcohol and water? What is a "constant boiling mixture?" Give four examples from this chapter and one from a much earlier chapter. (Hint: use the index.)
2. Why should normal hexyl alcohol be less soluble in water than isopropyl alcohol?
4. Approximately 2.6 gallons of ethyl alcohol (95% ethyl alcohol and 5% water) are produced per bushel of corn. Fermentation costs about 50 cents per gallon of ethyl alcohol and corn is selling for about $2.50 per bushel. If the operating costs of an ethyl alcohol plant are 30 cents per gallon, what is the maximum price that could be paid for ethylene to remain competitive with "natural" ethyl alcohol?

FORMALDEHYDE AND ACETALDEHYDE

"A precedent embalms
a principle."

■

Benjamin Disraeli, 1804–1881

A good warm-up to the discussion of aldehydes is a repeat and an elaboration of Chapter I's discussion of oxidation. The chemical route to many petrochemicals is oxidation—the reaction of an atom or a molecule with oxygen. The oxygen can come from air or from another compound that readily gives up its oxygen, such as hydrogen peroxide, H_2O_2.

If oxidation is taken to its extreme, that is, complete oxidation of an organic compound, you end up with only oxygen attached to each carbon atom (CO_2) and with water (H_2O). Burning (combustion) is an example of complete oxidation. Your body functions also are a good example. For instance, if vodka (ethyl alcohol) is ingested, the ultimate result is a chemical imbalance in your system:

$$C_2H_5OH + 3O_2 \longrightarrow 2CO_2 + 3H_2O.$$

The more vodka, then later on the more CO_2 and the less oxygen in the body. That's why you can partially relieve a hangover by breathing from an oxygen mask. It restores the normal oxygen balance to your body (and head).

In petrochemicals, partial oxidation, rather than complete, is more desirable and gives rise to several major classes of compounds.

Class	Formula	
Alkanes	$R\text{-}CH_3$	Increasing oxidation state ↓
Alcohols	$R\text{-}CH_2\text{-}OH$	
Aldehydes	$R\text{-}\overset{\overset{\displaystyle O}{\parallel}}{C}\text{-}H$	
Acid	$R\overset{\overset{\displaystyle O}{\parallel}}{C}\text{-}OH$	
Carbon dioxide	$CO_2 + H_2O$	

The aldehydes, specifically formaldehyde and acetaldehyde, are midway in this spectrum.

The aldehyde signature, $\text{-}\overset{\overset{\displaystyle O}{\parallel}}{C}\text{-}H$ (written also as -CHO, but never -COH), is always located at the end of the carbon chain. Common names for aldehydes are derived from the corresponding acid to which they are converted by further oxidation. The ending -ic acid is simply changed to -aldehyde:

Acid	Aldehyde
formic acid (H-COOH)	formaldehyde (H-CHO)
acetic acid (CH_3-COOH)	acetaldehyde (CH_3-CHO)
propionic acid (CH_3CH_2-COOH)	propionaldehyde (CH_3CH_2-CHO)

FORMALDEHYDE

Formaldehyde is the first member of the aldehyde family, and it illustrates how precedent can be the predicate to principle. Formaldehyde's public image has always been associated with the funeral homes, doctor offices, and biology classes as an embalming fluid, a disinfectant, and a preservative. Even before 1900, it was produced by the oxidation of methanol, which at the time was strictly wood alcohol coming from the destructive distillation of wood. In 1905, a major breakthrough in the technology of plastics was made by Dr. Leo Baekeland in Yonkers, New York. He found a process for producing a stable, cross-linked polymer—later named Bakelite after him. The ingredients were phenol and formaldehyde. By the 1920s, the growth of this resin strained the wood alcohol producing capacity, but the development of the methane reforming route to methanol relieved the situation. Formaldehyde technology has remained unchanged even to today, despite the volume growth which puts it in the top 25 commodity chemicals.

The Process

The commercial process has always been to react methanol and air in the presence of a catalyst. Recent processes have switched from metal to metal oxide catalysts, especially iron oxide and molybdenum oxide.

$$\underset{\text{methanol}}{CH_3OH} + \tfrac{1}{2} O_2 \longrightarrow \underset{\text{formaldehyde}}{H\text{-}CHO} + H_2O$$

The reactor in the formaldehyde plant shown in Figure 14–1 is really just a heat exchanger—a large vessel that contains a bundle of tubes through which the methanol and oxygen are pumped. Outside the tubes, but inside the vessel, is a liquid that is used to transfer the heat. But the reaction is exothermic. So even though the reaction must take place at 575–700°F, once it starts, the exchanger must take heat away from the methanol + oxygen-to-formaldehyde reaction, rather than provide the heat to make it work. The heat transfer liquid outside the tubes is continuously vaporized to take the heat away.

The tubes of the heat exchanger are filled loosely with the catalyst necessary for the reaction. Actual reaction time and residence in the tubes is less than one second.

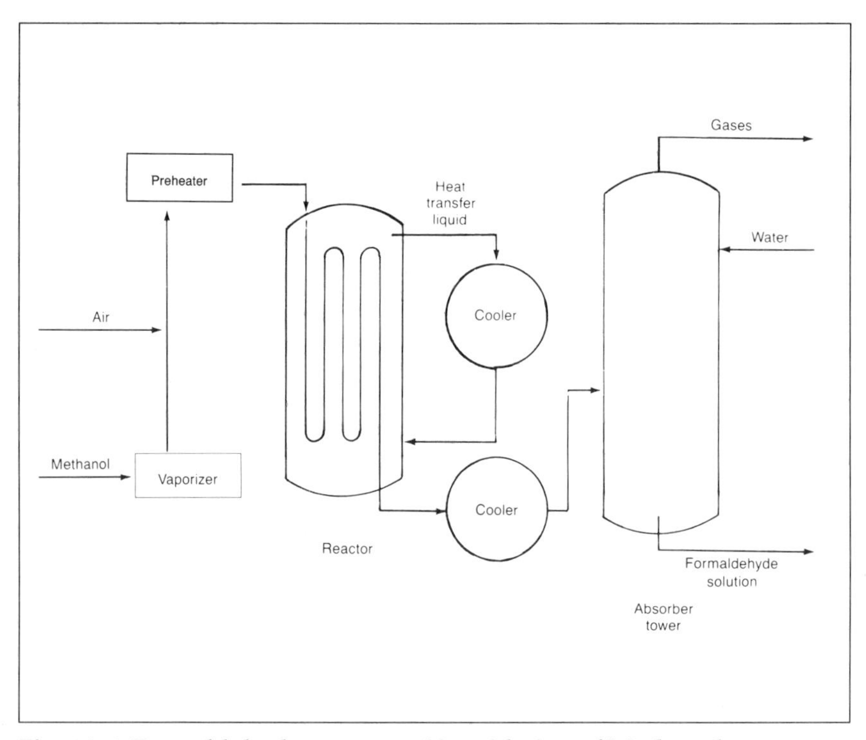

Fig. 14–1 Formaldehyde process—Air oxidation of Methanol

The hot reactor effluent gases are cooled and passed into a water absorption tower. Formaldehyde is water soluble and is separated from the remaining gases which exit the column overhead. Formaldehyde concentration in the tower is adjusted by controlling the amount of water added to the top of the tower. Generally, a product containing 37 to 56% formaldehyde in water is made. Methanol is often added as a stabilizer.

One of the major advantages of the metal oxide catalyst over that of the straight metal catalyst is the elimination of the need for a methanol recovery tower. The metal oxide catalysts result in not only high yields, but very high conversion rates. Consequently, there is no need to recover the small amounts of methanol that remain unreacted. It becomes part of the aqueous formaldehyde solution and serves as a stabilizer for the system. By-products are CO, CO_2, dimethyl ether and formic acid. The process yields (the percent of the methanol that ends up in formaldehyde) are 95 to 98%.

Material Balance	
Feed:	
Methanol	415 lbs.
Air (oxygen)	209 lbs.
Products:	
Formaldehyde	370 lbs.
Water	221 lbs.
By-products	33 lbs.

Special Properties and Handling Characteristics

Formaldehyde is a colorless, toxic gas at room temperature, with a pungent, irritating odor. It is flammable and explosive in presence of air. Both gaseous and liquid forms of formaldehyde polymerize at room temperature and because of this, it can only be maintained in the pure state for a very short period. Because of these unhandy conditions there are two ways formaldehyde gets into commerce, as water solutions of formaldehyde called formalin and as a solid called paraformaldehyde or trioxane.

Formalin concentrations range from 37 to 56% by weight formaldehyde, the balance water, with or without stabilizer. The market trend is to higher concentrations. In fact, over half of formalin merchant sales is shipped out as 50% or stronger solutions. The higher concentrations generally require stabilizers. No hazardous shipping label is required.

Paraformaldehyde is actually a polymer of formaldehyde that can easily be formed by removing water from a 50% formalin solution under reduced pressure. As the formaldehyde concentration increases, crystals of paraformaldehyde from spontaneously. It is available at 91–97% purity. It is more stable than neat formaldehyde but just as useful in applications, where it readily decomposes back to the straight stuff.

Trioxane is a cyclic trimer (three formaldehydes in a ring). It is formed by distilling 56% formalin in the presence of sulfuric acid. At 99% purity, it has a melting point of 147 °F and boiling point of 238 °F. It is soluble in water and the more common organic liquids. Trioxane is relatively stable and can be distilled without decomposition. But in the presence of a strong acid like hydrochloric or sulfuric, it easily depolymerizes to gaseous formaldehyde. This characteristic enables an immediate and controllable source of gaseous product for chemical reactions. No hazardous shipping labels are required for either grade.

Formaldehyde Properties	
Freezing point	−134.0°F (−92.0°C)
Boiling point	−5.8°F (−21.0°C)
Specific gravity	0.815 (lighter than water)

Commercial Aspects

More than 50% of the formaldehyde produced in the U.S. goes to making synthetic resins, primarily urea, melamine, and phenol-formaldehyde resins. These resins find multiple uses and applications as adhesives and insulation in housing construction and molded parts for the automotive, furniture, electronic, and appliance industries. Second in importance is use as an intermediate in the production of butanediol, which is used as a solvent, humectant, and plasticizers; in the production of tetrahydrofurane also used as a solvent, adhesive chemical intermediate, and printing inks component; and in the production of pentaerythritol (synthetic lubricants, varnishes) and hexamethylene diamine (Nylon 66 and pharmaceuticals). That leaves the smaller, older uses—fertilizer, germicide, disinfectant, embalming fluid, preservative. And since neatness and completeness count, you need to know that formaldehyde is used as a textile sizing agent and accounts for the nice fresh look in your newer clothes.

ACETALDEHYDE

Acetaldehyde is old. It is not ancient like ethyl alcohol, the essential ingredient in wine, but it owes its discovery to this closely related compound. Acetaldehyde was first prepared by Scheele in 1774 by dehydrogenation of ethyl alcohol. Just as many nicknames get attached to people at infancy, this process generated the name "aldehyde." It is a contraction for compounds that are alcohol dehydrogenates.

The close chemical relationship between acetaldehyde and ethyl alcohol is apparent in the grape fermentation process. The sugar in the grapes turn to acetaldehyde as an intermediate step. Fortunately for wine makers and oenophiles, the acetaldehyde immediately reduces to ethyl alcohol.

Oxidation of ethyl alcohol was one of the two important commercial routes to acetaldehyde until the 1950s. The other was the hydration of acetylene. The chemical industry was always after replacement of acetylene chemistry, not just for acetaldehyde production, but all its many applications. Acetylene was expensive to produce, and with its reactive, explosive nature, it was difficult to handle. In the 1950s, acetylene chemistry was phased out by the introduction of the liquid phase direct oxidation of ethylene. That also began the decline of the ethyl alcohol route as well. Presently almost all the acetaldehyde uses the newer process.

$CH_2=CH_2$	+	O_2	$\longrightarrow$	$CH_3\text{-}CHO$
ethylene		oxygen		acetaldehyde

The Process

The catalyst is the key to this reaction and in this case is an aqueous solution of palladium chloride ($PdCl_2$) and cupric chloride ($CuCl_2$). There is a complex, but well understood, mad scramble of ions and molecules that takes place as chlorine temporarily separates from the palladium and the copper and facilitates ethylene's reacting with oxygen.

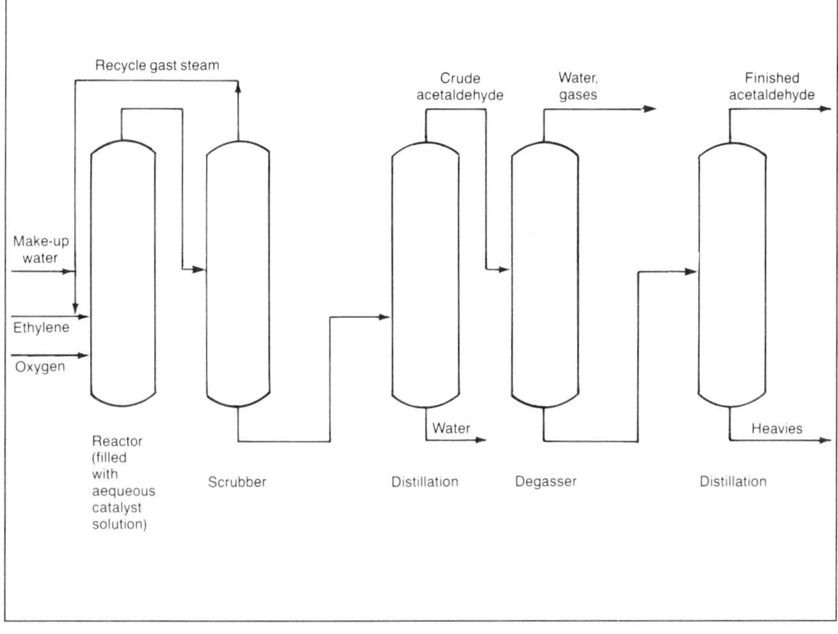

Fig. 14–2 Oxidation of ethylene to acetaldehyde

The process operates continuously and is easily regulated by the flow of fresh ethylene and oxygen to the reactor and the removal of acetaldehyde as a vapor.

As shown in Figure 14–2, high purity ethylene (99.7%) and oxygen (99.0%) are fed under pressure (100 psi) to a vertical reactor containing the aqueous catalyst solution. Reaction temperature is maintained at 250–275 °F. Because the reaction is exothermic, heat liberated is partially removed by vaporizing the water present in the reactor. Makeup water is continuously fed to the reactor to maintain proper catalytic solution concentration.

The gaseous reaction mixture containing steam, unreacted ethylene, and acetaldehyde are passed into a water scrubber where the acetaldehyde is dissolved in water and then removed and fed to a distillation column. The overhead gases from the scrubber are recycled to the reactor. In the distillation step, acetaldehyde is separated from water and a few cats and dogs (by-products). A final distillation step gives acetaldehyde in 99% purity.

Material Balance	
Feed:	
Ethylene	670 lbs.
Oxygen	404 lbs.
Catalyst	Small
Product:	
Acetaldehyde	1000 lbs.
By-products	74 lbs.

Commercial Aspects

Acetaldehyde is a colorless, flammable liquid with a pungent, fruity-like odor and having a boiling point close to room temperature (70 °F). It is soluble in water and most common organic solvents. It is a toxic chemical requiring care in handling. Acetaldehyde is commercially available as technical grade, 99% minimum purity. Because of its low boiling temperature, acetaldehyde must be contained in pressure vessels, including rail cars and tank trucks, which during a shipment, fly the hazardous material placard.

Acetaldehyde Properties	
Freezing point	−190.3 °F (−123.5 °C)
Boiling point	69.6 °F (20.9 °C)
Specific gravity	0.779 (lighter than water)
Weight per gal.	6.49 lbs./gal.

As early as World War I, acetaldehyde was the primary route to acetic acid and acetone. While other preferred technologies for acetone have been developed, acetaldehyde remains an important intermediate to acetic acid as well as several other chemicals.

Product	End Uses
Acetic acid and anhydride	Vinyl acetate, textile processing, cellulose acetate (cigarette filters) and aspirin
Pentaerythritol	Alkyd resins, explosives, synthetic lubricants, coatings
C_4 alcohols	Solvent, plasticizer, chemical intermediate
2-ethyl hexyl alcohol	Plasticizer for PVC

These four derivatives of acetaldehyde account for more than 80% of its total U.S. production.

Chapter XIV in a nutshell...

Aldehydes have the characteristic -CHO group, sometimes written as

$$-\overset{O}{\overset{\|}{C}}H$$

-CH. They are a dehydrogenated form of a corresponding alcohol. Formaldehyde, CH_2O corresponds to methanol, acetaldehyde, CH_3CHO, to ethyl alcohol.

The formaldehyde process is an air oxidation of methanol, CH_3OH, which has water as a by product. Formaldehyde is a gas at room temperature, but is usually handled either as a water solution called formalin or as polymers called paraformaldehyde and trioxane. Both are readily converted back to formaldehyde. Some uses of formaldehyde are the manufacture of polymer resins and as a germicide.

Acetalhyde is made by the direct oxidation of ethylene, C_2H_4. It is a liquid at room temperature and is an intermediate in the production of acetic acid, acetic anhydride, butyl and 2-ethyl hexyl alcohol.

Exercises

1. Fill in the blanks.

   a. An aqueous solution of formaldehyde. ____________
   b. Will polymerize. ____________
   c. Trioxane is a trimer of ____________ that easily reverts to ____________.
   d. A chemical intermediate. ____________
   e. Product of complete oxidation. ____________
   f. Dehydrogenation of alcohol. ____________
   g. Feedstock for acetic acid. ____________

2. What is an oenophile anyway, and why do they have an aversion to acetaldehyde?

3. Write out the oxidation chain that shows the progression from ethane to CO_2 + H_20, and includes the appropriate olefin, alcohol, aldehyde, and acid.

XV

THE KETONES (ACETONE, METHYL ETHYL KETONE, AND METHYL ISOBUTYL KETONE)

"Names are not always what they seem. The common Welsh name Bzjxxllwcp is pronounced Jackson."

■

Mark Twain, 1835–1910

There's no need for ill-humored comments about the ketones being a trio of pop singers. They're a family of organic compounds (first cousins to the aldehydes) that all have the ketone signature

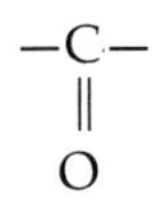

Somewhere in the middle of a hydrocarbon chain, the two hydrogens attached to a carbon are replaced by a double-bonded oxygen. Ketones come in many sizes and shapes. The convention for naming them is to refer to the radicals (alkyl groups) attached to the ketone signature. In Figure 15–1, the three commercially traded aliphatic ketones with the largest volumes are shown: acetone, MEK, and MIBK.

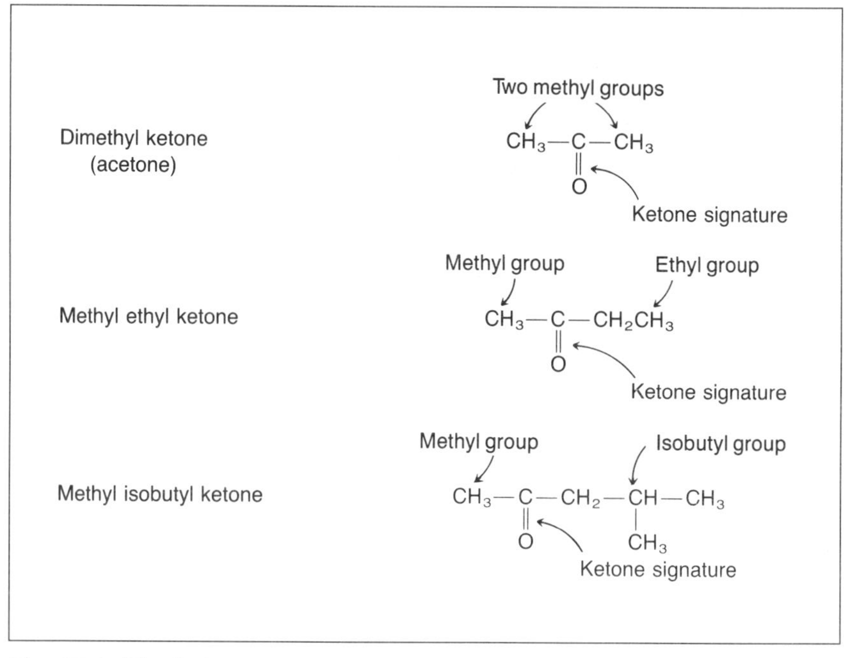

Fig. 15–1 The ketones

ACETONE

Right up front you need to know that acetone and dimethyl ketone (DMK) are the same thing chemically.

As far back as the pre-World War I years, acetone was used extensively as a solvent. The early commercial routes to producing acetone included the destructive distillation of wood, the fermentation of either starch or corn syrup, and the conversion of acetic acid. The development of olefin technology permitted a more efficient petrochemical route, the dehydrogenation of isopropyl alcohol, to replace the originals by the late 1930s. It remained the primary route through the 1960s.

One convenient way to classify today's processes for making acetone is to separate them into two categories, by-product and on-purpose. You'll recall (won't you?) that acetone is one of the outturns of the cumene-to-phenol process described in Chapter VII. (Approximately 0.6 pounds of acetone are generated for each pound of phenol.) That falls into the category of by-product production because the rate at which acetone is produced is not solely dependent on anticipated acetone demand. Often the demand for the phenol dictates the rates at which the phenol plant is run, and either the acetone is a disposal problem or a shortage exists. More than 70% of the acetone produced in the U.S. is in the "by-product" category.

The "swing" supply of acetone comes from the plants that produce acetone "on purpose" by catalytic dehydrogenation of IPA. There is nearly as much capacity in place in the U.S. to produce on-purpose acetone as there is via the phenol route. But on-purpose being the swing supply, the growing demand for phenol has resulted in by-product acetone supply shutting down most of the on-purpose capacity.

The Process

The primary on-purpose route to acetone looks like a typical petrochemical plant, if there is such a thing. Like a lot of other petrochemical processes, the "chemistry" part of the plant, the reactor, is simple. Together with all the other "mechanical" processes like heating, cooling, and especially separation that fill up the plant site, these make the acetone plant indistinguishable from many other plants.

The dehydrogenation route is shown in Figure 15–2. In this plant, the iso-propyl alcohol feed is heated to about 900°F in a preheater and then charged to a reactor at about 40–50 psi pressure. The reactor is filled with a catalyst of zinc oxide deposited on pumice. Pumice is a fine powder of silica dioxide, which is glass. It has many fine pores in which the catalyst can reside, and therefore has a very large surface area to expose to the IPA. The catalyst causes the hydrogen to pop off the OH group, forcing the double bond to the oxygen, the ketone signature.

The hot effluent from the reactor containing acetone, unreacted IPA, and hydrogen is cooled in a condenser and then scrubbed with water to remove the hydrogen. Both IPA and acetone are highly soluble in water, but hydrogen is not. So by washing the effluent with water, the hydrogen

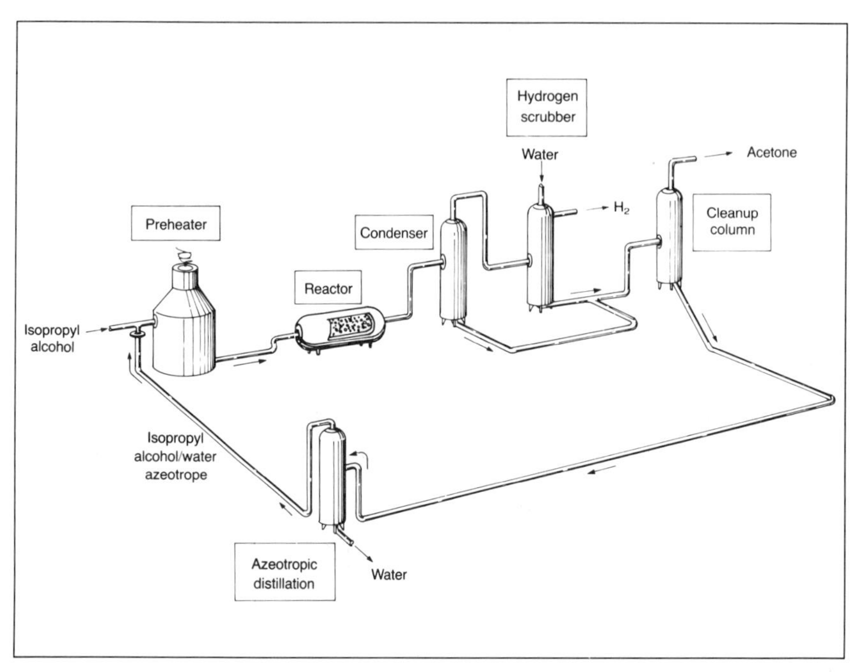

Fig. 15–2 IPA-to-acetone plant

bubbles out the top, and the IPA/acetone comes out the bottom with the water. The process is just like the solvent recovery scheme explained in Chapter II. The IPA/acetone/water is fractionated, with the acetone going overhead and the IPA/water coming out the bottom. The yield (the percent of the IPA that ends up in the acetone) is 85–90%.

The presence of the water causes special treatment of the IPA recycle stream. IPA and water will form an azeotropic solution, like the one discussed in Chapter III. The two compounds will boil together at a temperature different than the boiling point of either. Consequently, the stream recycled to the reactor contains about 9–10% water. The presence of water doesn't affect the IPA dehydrogenation step. It's just a little extra baggage, requiring a little more heating, cooling, and pumping around.

Material Balance	
Feed:	
Isopropyl alcohol	1158 lbs.
Product:	
Acetone	1000 lbs.
H_2	34 lbs.
By-products	124 lbs.

Other Routes

The more expedient, direct catalytic oxidation route to acetone was developed in Germany in the 1960s. If you had been in charge of building the acetone business from scratch, you'd probably not have built any IPA-to-acetone plants if you had known about the Wacker process. It's a catalytic oxidation of propylene at 200–250 °F and 125–200 psi over palladium chloride with a cupric (copper) chloride promoter. The yields are 92–94%. The hardware for the Wacker process is probably less than for the combined IPA/acetone plants. But once the latter plants were built, the economies of the Wacker process were not sufficient to shut them down and start all over. So the new technology never took hold in the U.S.

There are several other routes to acetone of minor importance: air oxidation of IPA; reaction between IPA and acrolein for the production of allyl alcohol, with acetone as the by-product; co-production when IPA is oxidized yielding acetone and H_2O_2, hydrogen peroxide, the principal ingredient of bleach; and by-product production from the manufacture of methyl ethyl ketone.

Commercial Aspects

Uses. Acetone is used in two basically different ways, as a chemical intermediate and as a solvent. As an intermediate, acetone is used to produce MIBK, methyl methacrylate (used to make plexiglass products), Bisphenol A (raw material for epoxy and polycarbonate resins), and higher molecular weight glycols and alcohols.

As a solvent, acetone is used in varnishes, lacquer, cellulose acetate fiber, cellulose nitrate (an explosive), and as a carrier solvent for acetylene in cylinders. Acetylene is stored at about 225 psi, but is so explosively reactive that as an extra precaution the cylinder is filled with asbestos wool soaked in acetone. Acetylene is extremely soluble in acetone, and the asbestos keeps it from sloshing around when the cylinder is half empty. Acetone also is used in smaller volumes for the manufacture of pharmaceuticals and chloroform (the anesthetic).

Properties and Handling. Acetone is a mobile, colorless, volatile, highly flammable liquid. It has an odor that makes you think you're in a hospital. Acetone dissolves in water, alcohol, ether, and most other organic solvents. That's why it's usually included in paint brush cleaner. It dissolves almost anything, and then can be washed away with water.

Acetone Properties	
Freezing point	−139.6°F (−95.4°C)
Boiling point	133.0°F (56.1°C)
Specific gravity	0.7901 (lighter than water)
Weight per gallon	6.6 lbs./gal.

Acetone is sold commercially in three grades, USP (99%), CP (99.5%), and technical (99.5%). The terms USP and CP are acronyms used in the trade and stand for U.S. pure and chemical pure. Acetone is shipped in run of the mill tank trucks, tank cars, and in drums. The hazardous material shipping placard must be displayed for this highly flammable liquid.

METHYL ETHYL KETONE

Most of what you read in the previous section about acetone also applies to methyl ethyl ketone (MEK). The processes for making MEK can be broadly categorized into "by-product" and "on-purpose"; the more popular processes are the same—they just start with larger molecules, and the applications are much the same.

The three popular manufacturing routes to MEK are:

1. dehydrogenation of secondary butyl alcohol (instead of IPA);
2. air oxidation of butylene (instead of propylene);
3. catalytic oxidation of butane to form acetic acid and by-product MEK.

In the first route, not only is there close similarity in the chemistry of the acetone/dehydrogenation route, the hardware is almost identical to the plant shown in Figure 15–2. The reaction is as follows:

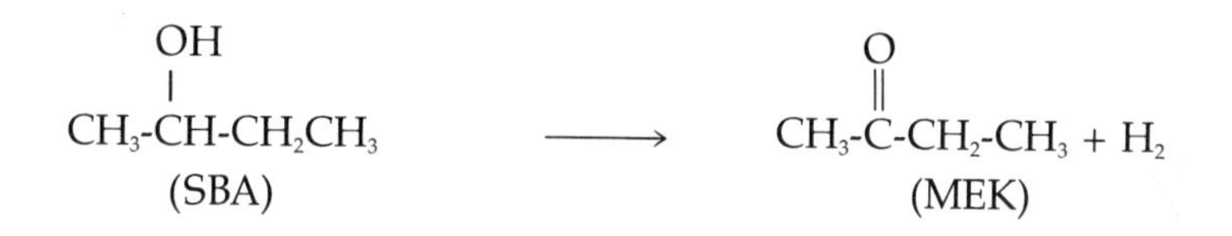

The heated secondary butyl alcohol (SBA) vapors are passed through a reactor containing zinc oxide catalyst at 750–1000°F at atmospheric pressure. The catalyst causes the hydrogen to pop off, forming MEK. The separation of the reactor effluent into MEK, water, hydrogen, and recycle SBA is about the same as Figure 15–2. The overall yield (the percent of SBA that ends up as MEK) is about 85–90%.

There is a more energy-efficient version of this process that takes place in the liquid phase. A catalyst of very fine Raney nickel or copper chromate, suspended in a heavy, high boiling temperature solvent, is mixed with SBA. At 300–325°F the SBA undergoes dehydrogenation to MEK. As it does, the MEK and hydrogen immediately vaporize, leaving the reaction medium in gaseous form, and need only to be separated from each other.

The second, on-purpose route to MEK is the direct oxidation of butylene, the Wacker process:

$$\underset{\text{Butylene}}{3CH_2{=}CH\text{-}CH_2\text{-}CH_3} + 3O_2 \longrightarrow \underset{\text{MEK}}{2CH_3\text{-}\overset{\overset{\displaystyle O}{\|}}{C}\text{-}CH_2\text{-}CH_3} + 4H_2O$$

With reaction conditions of 200–225°F, 150–225 psi, and a palladium chloride-cupric chloride catalyst, MEK yields are 80–90%. The operating costs of the Wacker process for MEK (and acetone and several other petrochemicals as well) are relatively low. But the plant is made of more expensive materials. Because of the corrosive nature of the catalyst solution, critical vessels and the piping must be titanium (that's expensive!) based, and the reactor is rubber-lined, acid-resistant brick.

The third route, producing by-product MEK, accounts for only a modest portion of the total supply, less than 15%. Plants designed to produce acetic acid from the direct oxidation of butane can be run to produce almost no MEK. But optimum operating cost balanced against market product prices usually warrants shifting to a 60/40 acetic acid/MEK outturn.

Material Balance

Feed:	
Secondary butyl alcohol	1140 lbs
Product:	
Methyl ethyl ketone	1000 lbs.
Hydrogen	28 lbs.
By-products	112 lbs.

Commercial Aspects

MEK is used in a variety of ways as a solvent. It owes part of its popularity to the fact that it is considered non-toxic, and therefore not an air pollutant. As a result, it is used as a low boiling temperature solvent in coatings—vinyl, nitrocellulose, and acrylic to name a few. In these applications, MEK flashes (vaporizes or "dries") quickly at room temperatures, leaving the coating behind. Since the MEK ends up in the atmosphere, its non-polluting character is important.

MEK also is used as the solvent in lube oil dewaxing, wood pulping, and toluene (see Chapter III), and in the manufacture of printing ink.

The physical characteristics of MEK are similar to those of acetone. It's colorless, mobile, flammable, and sweet smelling, if that's what you call a hospital smell. It's very soluble in water and most common organic solvents. There are only two grades commercially traded, technical (99%) and CP (99.95%). Shipping and handling are similar to acetone.

Methyl Ethyl Ketone Properties	
Freezing point	−123.5°F (−86.4°C)
Boiling point	175.3°F (79.6°C)
Specific gravity	0.806 (lighter than water)
Weight per gallon	6.7 lbs./gal.

METHYL ISOBUTYL KETONE

MIBK is more complicated than the one step conversion process for acetone and MEK. Manufacture of MIBK takes the three step process shown in Figure 15–3, starting with acetone.

First, the acetone is condensed (or reacted or dimerized) with itself. That is, it's passed over a catalyst and two acetone molecules chemically react to form diacetone alcohol. Both the acetone signature and the alcohol signature (-OH) are sported by diacetone alcohol. The catalyst is an alkaline compound like $Ca(OH)_2$ (calcium hydroxide or soda lime), and the reaction is run at about 32°F.

In the second step, the diacetone alcohol is dehydrated (the -OH group and a hydrogen atom are clipped off) to form mesityl oxide. The dehydration is done by mixing the diacetone alcohol with a sulfuric acid catalyst at 212–250°F.

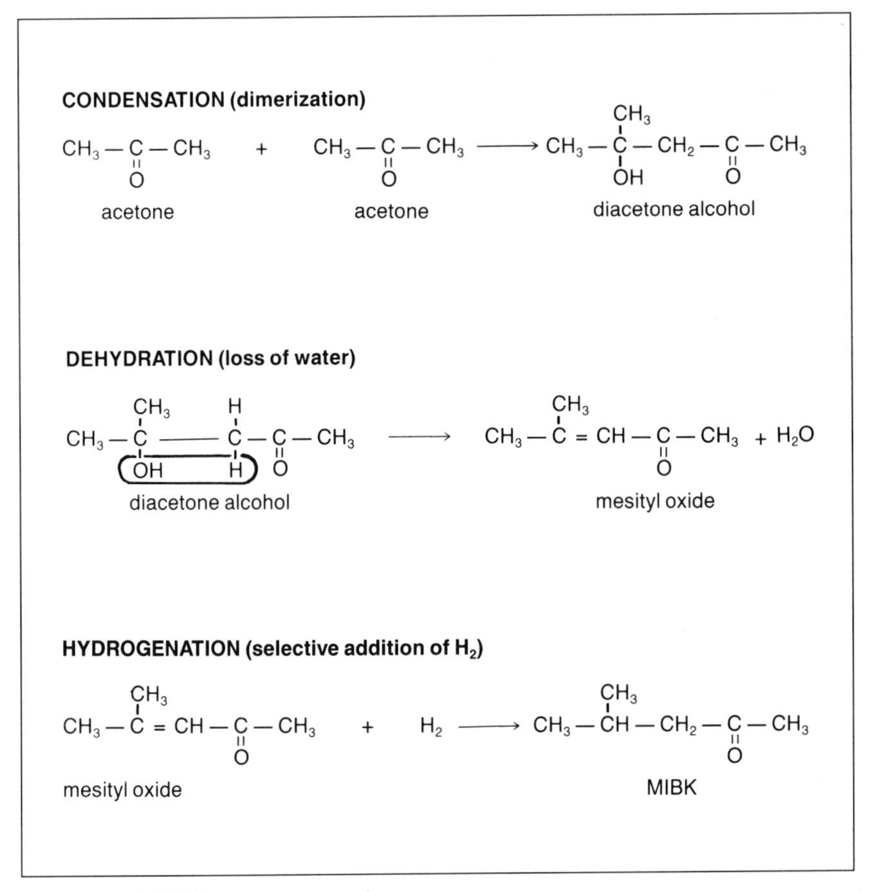

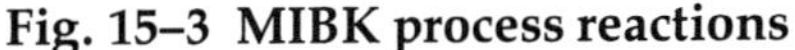

Fig. 15–3 MIBK process reactions

In the third step, the mesityl oxide is hydrogenated (hydrogen added) to MIBK by heating it to the vapor stage at 300–400 °F and passing it over a copper or nickel catalyst at 50–150 psi.

Material Balance	
Feed:	
Acetone	1160 lbs.
Hydrogen	20 lbs
Product:	
Methyl isobutyl ketone	1000 lbs.
Water	138 lbs.
By-products	44 lbs.

One problem with this process is the difficulty of controlling the last step. As the MIBK gets formed, it also has the tendency to hydrogenate further to methyl isobutyl carbinol. Further addition of hydrogen wipes out the ketone signature, replacing it with the hydroxyl group, -OH. This unavoidable by-product, methyl isobutyl carbinol, has to be separated from the MIBK by fractionation. The overall yield of MIBK (the amount of acetone that ends up as MIBK) is around 90%.

Alternate Processes

A small amount of MIBK is made from a new European-originated process. A complex catalyst system involving palladium metal and a cation exchange resin is used. The reaction permits going directly from acetone to MIBK.

A process that's similar in concept involves going directly from IPA to a mixture of acetone and MIBK. The process is confidential, and details aren't available at this writing.

Commercial Aspects

Uses. The applications of MIBK read a lot like those of MEK. In the 1960s and 1970s, MIBK rapidly replaced the use of ethyl acetate and butyl acetate as a solvent for resins. However, MEK is now a better competitor in many of these applications because MIBK is suspected of being a pollutant and is being replaced in many coating-solvent applications.

Some unique applications for MIBK include metallurgical extraction (particularly plutonium from uranium), a reaction solvent in pharmaceuticals, an adhesive, and, if you stretch the definition of application, the manufacture of methyl isobutyl carbinol.

Properties and Handling. MIBK is a colorless liquid and has a pleasant, almost fruity, odor. Unlike acetone and MEK, it is only slightly soluble in water. That happens to solvents as the size of the molecule gets larger. Most of the commercial trade in MIBK is in the technical grade (98.5%). Bulk shipments can be handled in conventional tank trucks and tank cars, but the hazardous material markings must be displayed.

MIBK Properties	
Freezing point	−119.0°F (−84.0°C)
Boiling point	241.0°F (116.0°C)
Specific gravity	0.8024 (lighter than water)
Weight per gallon	6.7 lbs./gal.

Chapter XV in a nutshell...

Ketones have the characteristic $-\overset{\overset{\displaystyle O}{\|}}{C}-$ signature group imbedded in them. Acetone, CH_3COCH_3, comes from two different routes. It is a by-product in the cumene to phenol/acetone process. It is the "on-purpose" product of the catalytic dehydrogenation of isopropyl alcohol. Acetone is popular as a solvent and as a chemical intermediate for the manufacture of MIBK, methyl methacrylate, and Bisphenol A.

Methyl ethyl ketone (MEK) and methyl isobutyl ketone (MIBK) have higher boiling temperatures, are less hazardous liquids, and are also popular as solvents. MEK is made by dehydrogenation of secondary butyl alcohol or the direct oxidation of butylene. MIBK is made via a three step process starting with acetone.

Exercises

1. Why are aldehydes and ketones "first cousins"? What are the signature groups of the two?
2. MIBK competes in the coatings solvent market with ethyl acetate. Because of its superior properties, MIBK can sell at a premium. When ethyl acetate is selling at 40 cents/lb., MIBK can fetch 45 cents/lb.

   The Skim Chemical Company can rent some IPA plant, acetone plant, and MIBK plant capacity from the Takesits Toll Company for 5 cents/lb. of product. When ethyl acetate is 40 cents/lb., what is the breakeven price that Skim can pay for propylene if the by-products of each plant are worthless but hydrogen costs 20 cents/lb.?
3. Fill in the blanks:

   a. ____________ Used as an anesthetic.
   b. ____________ Feedstock for MIBK.
   c. ____________ Will polymerize.
   d. ____________ Used as a solvent.
   e. ____________ Hydrogenation catalyst.
   f. ____________ Dehydrogenation catalyst.
   g. ____________ Feedstock for acetone.
   h. ____________ Direct oxidation of propylene.
   i. ____________ Major chemical use for acetone.
   j. ____________ Feedstock for MEK.
   k. ____________ Lube oil dewaxing agent.
   l. ____________ By-product in MIBK production.
   m. ____________ Oxidation catalyst.

   isopropyl alcohol
   acetone
   chloroform
   MIBK
   methyl methacrylate
   palladium chloride
   ether
   zinc oxide
   nickel
   propylene
   MEK
   methyl isobutyl carbinol
   Wacker process
   secondary butyl alcohol
   ethylene
   formaldehyde
   butylene

XVI

THE ACIDS

"Eye of a newt and
toe of a frog,
Wool of a bat and
tongue of a dog."

■

MacBeth
Shakespeare, 1564–1616

There are dozens of organic acids that are used in petrochemicals processing. But there are three that account for more than 70% of the total volume—acetic acid, adipic acid, and the phthalic acids. These compounds have little in common with each other besides the carboxyl signature group, written as -COOH, drawn as

$$-\underset{\underset{\displaystyle O}{\|}}{C}\text{-OH}$$

The group is so-called because it's a combination of the carbonyl (-C=O) and hydroxyl (-OH) groups.

You can think of carboxylic acids as being third down the line on the route to oxidizing a paraffin completely.

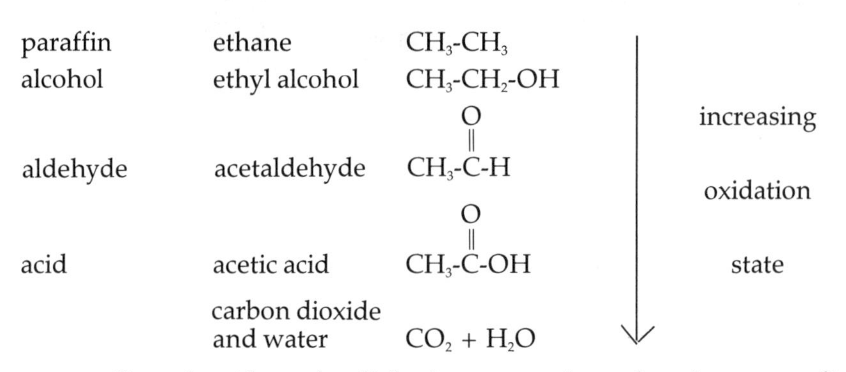

Organic acids can be aliphatic or aromatic, and can be mono-, di-, or polycarboxylic. Aliphatic acids have paraffinic hydrocarbon chains as their roots. The higher molecular weight aliphatic acids, the ones with greater than 12 carbons, are often referred to as the fatty acids because many of them were originally obtained by the hydrolysis of animal fat or vegetable oil. The word aliphatic is from the Greek aleiphatos, meaning fat.

The aromatic acids, as you would suspect from the name, have a benzene ring connected directly to the carboxyl signature group. Dicarboxylic acids have carboxyl groups attached in two places. Monocarboxylic acids have only one, and of course, the poly acids have three or more.

ACETIC ACID

Acetic acid is one of the simplest members of the aliphatic acid family. It has a methyl group attached to the acid signature group: CH_3COOH. Acetic acid is easily the largest volume organic acid produced in the U.S. (over three billion pounds per year). There are several places you're likely to find acetic acid or its derivatives. Acetic acid is the natural component of vinegar that gives it the characteristic smell. (Acetum is the Latin word for vinegar.) Acetic acid also is used to make acetates, which are polymerized and processed into adhesives and water base paints (polyvinyl acetate) and fibers (cellulose acetate, like Arnel).*

Manufacturing Acetic Acid

The destructive distillation of wood to produce methanol results in some by-product acetic acid, and that was the most popular but now defunct commercial source. The oldest, indeed the ancient method, fermentation, is still used to produce vinegar for the food industry. Vinegar is a 3–5% solution of acetic acid in water.

Most of the "on-purpose" acetic acid is made by one of the following routes:

Oxidation of acetaldehyde

$$CH_3CHO + 1/2\,O_2 \longrightarrow CH_3COOH$$

Oxidation of butane

$$CH_3CH_2CH_2CH_3 + O_2 \longrightarrow CH_3COOH + \text{by-products}$$

Carbonylation of methanol

$$CH_3OH + CO \longrightarrow CH_3COOH$$

Table 16–1 summarizes the vital statistics of the various routes. The methanol route is the latest and the state of the art technology and is now the preferred route to acetic acid. Over 65% of the acetic acid in the U.S. is from carbonylation of methanol.

Table 16–1 Acetic Acid Processes

Process	Catalyst	Reaction °F	Pressure PSI	Yield %	By-product
Butane oxidation	Cobalt acetate	300–450	800	57	Acetaldehyde acetone methanol
Acetaldehyde oxidation	Manganese acetate	150	0	95	none
Methanol carbonylation	Rhodium iodide	350–475	200	99	none

*Acetates are esters of acetic acid. Remember from Chapter I that an ester has a signature group

$$\text{-}\underset{\underset{O}{\|}}{C}\text{-O-R},$$ and the ester's

name usually comes from the acid and ends in the suffix -ate, as does the word acetate itself. The acetate group is made from acetic acid by replacing the carboxyl hydrogen with a R group:

$$CH_3\text{-}\underset{\underset{O}{\|}}{C}\text{-O-R}.$$

If the R is the vinyl group $-CH=CH_2$, for example, you have vinyl acetate,

$$CH_3\text{-}\underset{\underset{O}{\|}}{C}\text{-O-}CH=CH_2.$$

If the R group is the ethyl group $-C_2H_5$, then you have ethyl acetate,

$$CH_3\text{-}\underset{\underset{O}{\|}}{C}\text{-O-}C_2H_5.$$

Acetic Acid Plants

Acetaldehyde process. A stainless steel, water jacketed kettle is filled with concentrated (99%) acetaldehyde and catalyst. Then air is bubbled through for about 12 hours, where it reacts with the acetaldehyde. The gases exiting the kettle, still mostly air, are bubbled through water to "scrub" them and then discharged to the atmosphere. Some of the acetaldehyde ends up in the scrubbing water, but this is recovered by distillation.

The mixture in the kettle, which is crude acetic acid, is distilled to glacial acetic acid, 99% purity.

Butane process. Butane is added to a vessel that contains a solution of acetic acid and the cobalt acetate catalyst. The pressure is kept high enough to keep the butane liquid. Air is then bubbled through at 300–450°F and mixed vigorously. The volatile by-products come out the top—methane, carbon dioxide, and unreacted air. Crude acetic acid is drawn off the bottom and distilled to give glacial acetic acid.

Methanol. BASF introduced high pressure technology in 1960 to make acetic acid out of methanol and carbon monoxide instead of ethylene. Monsanto subsequently improved the process by catalysis, permitting operations at much lower pressures and temperatures. The methanol and carbon monoxide, of course, come from a synthesis gas plant.

The reaction is run at about 350°F and 200 psi, with a rhodium iodide catalyst and sodium iodide promoter. The yield of acetic acid is about 99%. That is, about 99% of the methanol that gets converted ends up as acetic acid. By-products include only small amounts of dimethyl ether and methyl acetate.

Conversion rates as high as 99% are not encountered very often in the petrochemical industry. That, coupled with relatively mild operating conditions made this route the economic favorite since it was introduced.

Material Balance

Feed:	
Methanol	540 lbs.
Carbon monoxide	473 lbs.
Catalyst	small
Product:	
Acetic acid	1000 lbs.
By-products	13 lbs.

Commercial Aspects

Three quarters of the acetic acid produced in the U.S. is used to make vinyl acetate and cellulose acetate. Nearly all the vinyl acetate ends up as polyvinyl acetate, used to make plastics, latex paints, and adhesives. Cellulose acetate is predominantly a textile yarn but also is the white stuff in cigarette filters. It is used in the manufacture of plastic sheeting and film and in formulating lacquers.

Acetic acid also finds use as a chemical intermediate in the production of acetate esters for paint solvents and as a reaction solvent for the manufacture of terephthalic acid. Also acetic acid is the source of the acetyl group in the manufacture of acetyl salacylic acid.

Acetic Acid Properties

Boiling point	244.6°F (118.1°C)
Freeze point	61.9°F (16.6°C)
Specific gravity	1.0492 (heavier than water)
Weight per gallon	8.64 lbs./gal.

Properties and Handling

Acetic acid has the strong, pungent odor of vinegar. It is a colorless liquid that is soluble in water and most organic solvents. The concentrations of the commercial grades vary all over, as low as 3%. The USP glacial acetic acid is 99.5% pure.

Since acetic melts at 62°F, shipping pure grades poses a special problem. Cool weather can cause freezing, expansion, and container rupture.

Tank cars and trucks must be specially lined because of the reactive nature of the acid. Even dilute acetic acid will react if left long enough, as any chef who has made sauerbraten will attest. That's how he got that tough piece of chuck steak so tender—by reacting (soaking) it in vinegar, among other things. Culinary considerations aside, the white hazardous (corrosive) shipping label is required for concentrated acetic acid.

ADIPIC ACID

You can use analogies to put adipic acid in its right place. Acetic acid is the most important aliphatic monocarboxylic acid; adipic is the most important aliphatic dicarboxylic acid. (You remember, of course, that carboxylic is the contraction for carbonyl and hydroxyl, -C=O and -OH, or together, -COOH. Right?) Also, adipic acid is to Nylon 66 what cumene is to phenol. About 95% of the adipic acid ends up as Nylon 66, which is used for tire cord, fibers, and engineering plastics.

Adipic acid is produced by oxidizing cyclohexane. The two-step process shown in Figure 16–1 is used for almost all production. Cyclohexane is oxidized with air over a cobalt naphthenate catalyst to give a mixture of cyclohexanol and cyclohexanone. These two products are separated from the unreacted cyclohexane and then hit with a 50% nitric acid solution. That opens up the C_6 ring and adipic acid is formed. Yields are in the 90–95% range.

An alternate route to the cyclohexanone is sometimes used—hydrogenating phenol. The rest of the process is then the same, except that yields are in the 70% range. Since the by-product yield is so high, the process has had limited acceptance, with only about 5% of the adipic acid being made this way.

Material Balance

Feed:	
Cyclohexane	625 lbs.
Nitric acid	excess
Air	470 lbs.
Catalyst	small
Product:	
Adipic acid	1000 lbs.
$NO + NO_2$	small
N_2O, N_2, and air	residual
By-products	77 lbs.

Commercial Aspects

Since almost all adipic acid is used for Nylon 66, the primary producers are Nylon 66 manufacturers.

The fibers made from Nylon 66 are durable, tough, and abrasion resistant, which suits them for tire cord. They are easy to color, which gives them a secure place in the carpet market (and on the floor). The additional attributes of moldability or processability make Nylon 66 suitable in the engineering plastics market.

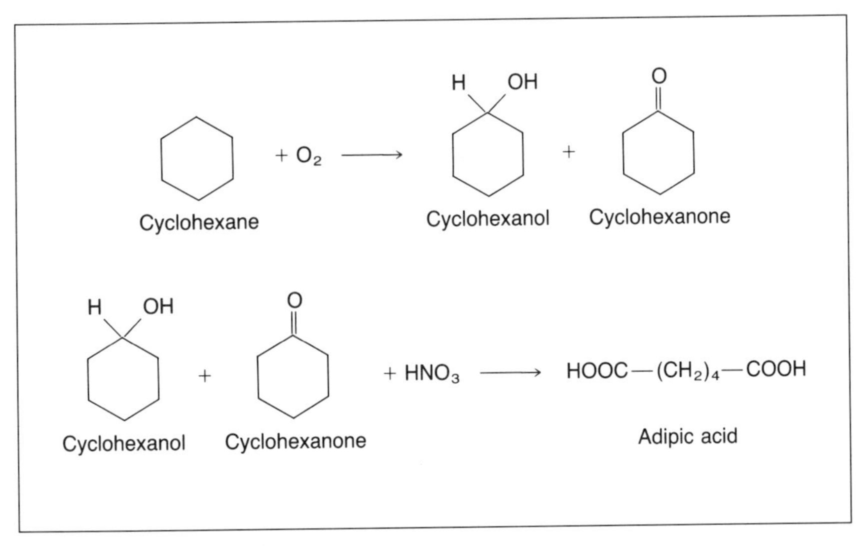

Fig. 16–1 Adipic acid process

Properties and Handling

Adipic doesn't physically fit the usual image of an acid. Its melting temperature is 306 °F. At normal temperatures, it is a white, crystalline powder that can be transported in one-ton cartons and in drums and 50 pound bags. Adipic acid is only slightly soluble in water but dissolves in alcohol. The commercially traded grade is 99.5% pure.

Adipic acid is an approved food additive and is one of the few solid petrochemicals manufactured on a commercial scale. (Terephthalic acid is another.)

PHTHALIC ACIDS

The strange spelling of these acids comes from the shortening of the original form, naphthalic acids. Naphtha originally came from an ancient Iranian word which was pronounced neft. It referred to a flammable liquid which oozed out of the earth. The word later anglicized to naphtha. Lexicography aside, the phthalic acids are made from the three xylenes, ortho-, meta-, and para-xylene, and are shown in Figure 16–2.

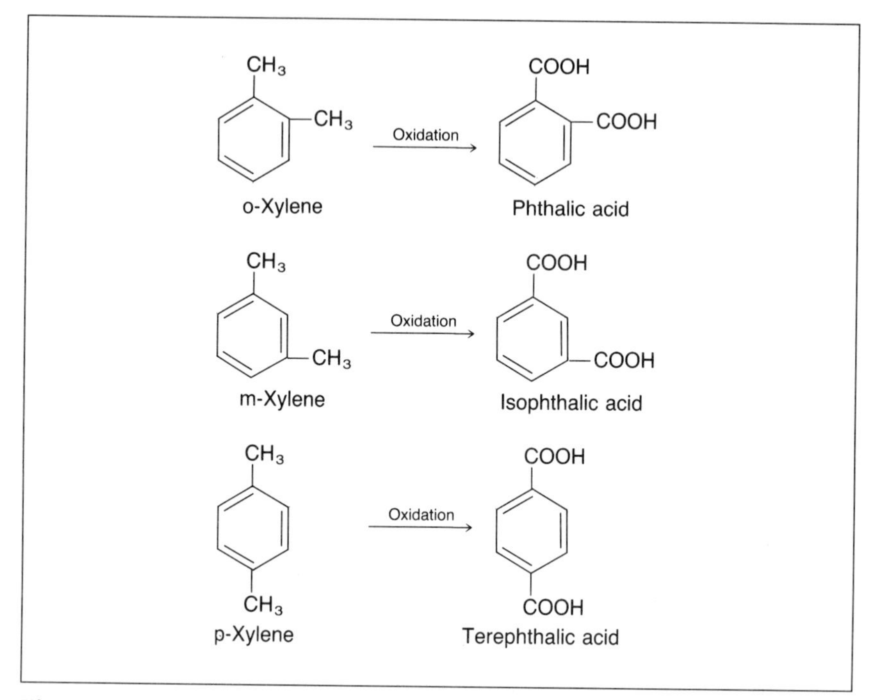

Fig. 16–2 Xylenes and phthalic acids

The major uses of these dicarboxylic acids (two -COOH groups in each) are plasticizers for polymers, alkyd and polyester resins, and fibers. All these applications are discussed in Chapters XX and XXI.

Phthalic Acid and Phthalic Anhydride

In the primary application of phthalic acid, life is rather transitory. Almost all phthalic acid is used to make phthalic anhydride. When ortho-xylene is used as the starting base chemical, phthalic acid is formed but immediately dehydrates (loses a molecule of water) to form phthalic anhydride, as shown in Figure 16–3.)

Until 1959, all the phthalic anhydride was made from coal tar naphthalene, the double-benzene ring compound also shown in Figure 16–3 that was easily oxidized directly to phthalic acid. But with phthalic anhydride being only a small share of coal oil, and with the demand for phthalic anhydride escalating rapidly, coal tar became an inadequate source. The frantic search for an alternative route led to the development of the recovery process for ortho-xylene from refinery aromatics streams discussed in Chapter III and the conversion of ortho-xylene to phthalic acid and anhydride. With the continued growth in the need for plasticizers and the inelasticity of naphthalene supply, ortho-xylene now accounts for 90% of the phthalic anhydride supply in the U.S.

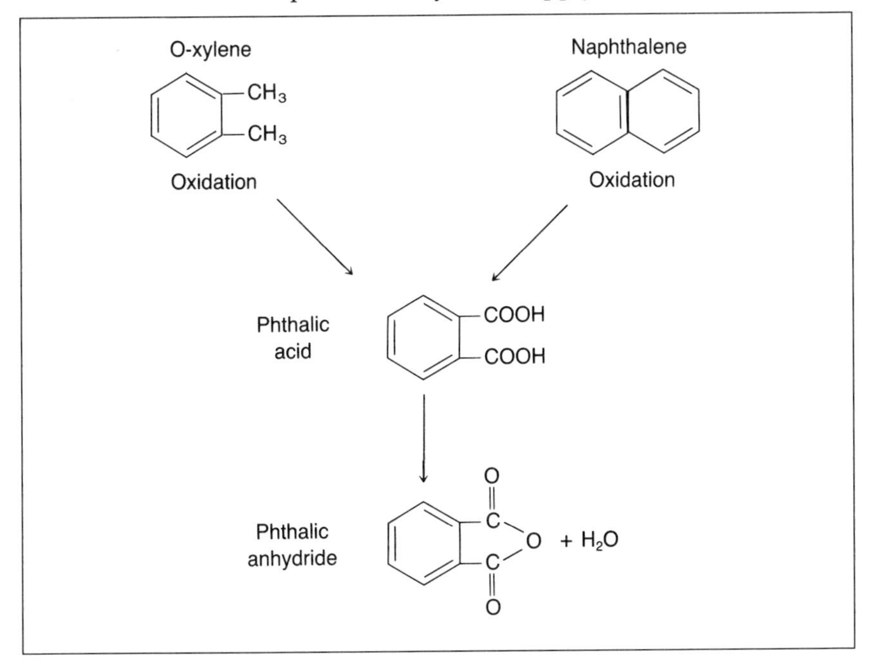

Fig. 16–3 Phthalic anhydride processes

The Process. A typical process for phthalic anhydride starts with mixing hot o-xylene vapor with excess preheated air. The gaseous mixture is then fed to a reactor consisting of tubes packed with vanadium pentoxide catalyst on a silica gel. Like most oxidation reactions, this one is exothermic, and the heat of reaction must be removed from the tubes to maintain the reaction temperature at about 1000°F. Contact time between the reactants and the catalyst is about a tenth of a second. The reaction gases—mainly phthalic anhydride, carbon dioxide, and water—are cooled, condensed, and purified in stainless steel facilities. Since phthalic anhydride solidifies at 269°F, the purifed (99.5%) product can then be flaked and packaged for shipping. Minor amounts of by-products, maleic anhydride, phthalic acid, and benzoic acid, are produced also.

Material Balance	
Feed:	
Ortho-xylene	975 lbs.
Excess air (7 times)	2061 lbs. (oxygen)
Product:	
Phthalic anhydride	1000 lbs.
Water	497 lbs.
By-products	210 lbs.
Unreacted oxygen	1329 lbs.

Properties and Handling. At ambient temperatures, phthalic anhydride is a white crystalline solid. It is slightly soluble in water. It is commercially available in two grades—pure (99.5%) and technical (99%). It is shipped in drums and bags in the solid form. Liquid phthalic anhydride is shipped in heated tank cars and trucks. It is not classified as a hazardous material because it is not corrosive or flammable.

Phthalic Anhydride Properties	
Melting point	269.0°F (131.6°C)
Boiling point	563.2°F (295.1°C)
Specific gravity	1.527 (heavier than water)
Physical appearance	white crystalline flakes or needles

Applications. Phthalic anhydride is used largely to make plasticizer for polyvinyl chloride. It is also a feed for alkyd resins and for polyesters which are widely used in construction, marine, and synthetic marble applications. Other minor applications are dyes, esters, and drying oil modifiers.

Terephthalic Acid

The sole use for para-xylene is to make terephthalic acid (TPA) and its derivative, dimethyl terephthalate (DMT). When DMT is copolymerized with ethylene glycol, chemists call it polyethylene terephthalate. On Seventh Avenue in New York they call it polyester. On the labels it is sometimes called Dacron.

The acid, TPA, is not unstable like phthalic acid. TPA can't dehydrate to the anhydride because the two acid groups, -COOH, aren't in the right places. So for the most part, TPA is the product that is traded commercially and DMT is produced en route to polyester.

The original route from p-xylene was oxidation in the presence of nitric acid. But the use of nitric acid is always problematical. There are corrosion and potential explosion problems, problems of nitrogen contamination of the product, and the problems due to the requirement to run the reactions at high temperatures. Just a lot of problems that all led to the development of the liquid air phase oxidation of p-xylene. Ironically the nitrogen contamination problem was the reason that the intermediate DMT route* to polyester was developed, since that was easy to purify by distillation. Subsequently, DMT has secured a firm place in the processing scheme.

The TPA Process. The newer technology, involves the oxidation of p-xylene, as shown already in Figure 16–2. The reaction takes place in an acetic acid solvent at 400°F and 200 psi, with a cobalt acetate/manganese acetate catalyst and sodium bromide promoter. The reaction time is about one hour. Yields are 90–95%, based on the amount of p-xylene that ends up as TPA. Solid TPA crystals drop out of solution as they form. They are continuously removed by filtration of a slipstream from the bottom of the reactor. The crude TPA is purified by aqueous methanol extraction which gives 99+ percent pure flakes.

Material Balance

Feed:	
Para-xylene	680 lbs.
Air (excess)	1843 lbs. (oxygen)
Catalyst	small
Product:	
Terephthalic acid	1000 lbs.
Water	230 lbs.
By-products	53 lbs.
Unreacted oxygen	1240 lbs.

Properties and Handling. Terephthalic acid has a high melting point, 572°F. At room temperature it is a white crystalline solid, insoluble

*High purity DMT is produced by esterification of TPA. That is, the terephthalic acid is reacted with methanol to form an ester (actually a di-ester),

$$CH_3O\text{-}\overset{O}{\overset{\|}{C}}\text{-}C_6H_4\text{-}\overset{O}{\overset{\|}{C}}\text{-}OCH_3$$

In the DMT process, the esterification is done by feeding a slurry of TPA crystals in methanol to a reactor with a catalyst of sulfuric acid at 220°F and 50 psi. DMT forms and can be purified by distillation. Yields exceed 95%, based on the TPA that ends up as DMT.
In some later designs resulting in less severe operating conditions, MEK or acetaldehyde have been used as promoters in place of sodium bromide.

in water or acetic acid. It is commercially available in fiber grade (99%) and technical grade (97%). Just to confuse things, the fiber grade of TPA is referred to as PTA. TPA and PTA are routinely shipped in bags, drums, and hopper cars as flakes. No hazardous shipping label is required.

Terephthalic Acid Properties	
Melting point	Sublimes* at 572.0°F (300°C)
Specific gravity	1.51 (heavier than water)
Physical appearance	White crystals or powder

*Goes directly from solid to vapor without passing through the liquid phase

Applications. About 95% of the TPA is used to make polyester. Most of that goes into fiber production, some into films (magnetic tapes, photographic materials, and electrical insulation). The route to fibers is through DMT or through a direct process if PTA is used. Minor amounts of TPA are used for herbicides, adhesives, printing inks, coatings, and paints. Polybutylene terephthalate is a molding resin used as an engineering plastic.

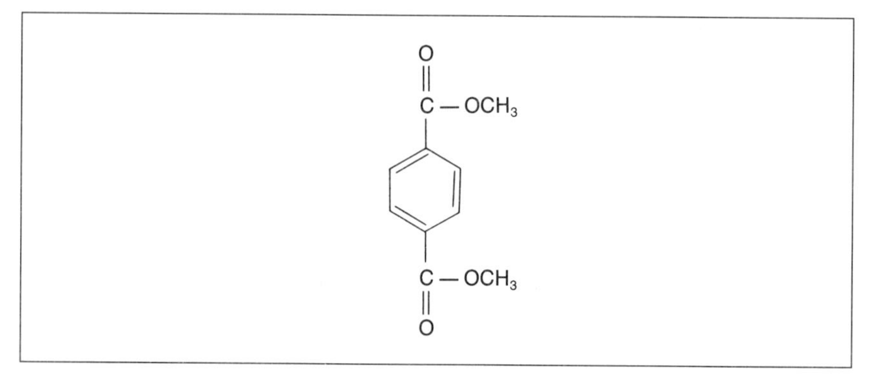

Fig. 16–4 Dimethyl terephthalate

Isophthalic Acid

The step-sister of the other two phthalic acids is isophthalic acid, made from meta-xylene. The applications are similar, but the commercial demand is smaller. If it weren't for the fact that m-xylene is a co-product of the other xylenes, no one would invent it. The other xylenes and phthalic acids would probably suffice.

Isophthalic acid is made by the same process as TPA, liquid phase air oxidation. Yields are about 80%. Isophthalic does have some unique redeeming value—it will enhance to some extent the mechanical and temperature sensitive properties of polyesters, alkyd resins, and glass reinforced plastics.

Chapter XVI in a nutshell...

Organic acids can be thought of as oxidation of corresponding alcohols, since they have the characteristic -OH signature group, plus a double bonded oxygen: -COOH or

$$\overset{\displaystyle O}{\overset{\|}{-C}}\text{-OH.}$$

Acetic acid, CH_3COOH, can be made by the oxidation of acetalhyde, CH_3CHO; by catalytic addition of CO to methanol; or by butane oxidation. Most acetic acid is used to make vinyl acetate or cellulose acetate, which are the intermediates for plastics, paints, adhesives, yarn, and cigarette filters.

Adipic acid is made by the reaction of nitric acid and cyclohexane. The adipic acid has two -COOH groups, which makes it very reactive. Adipic is used primarily for making Nylon 66.

The phthalic acids are made by oxidation of the corresponding xylene isomer. They are used for plasticizers and in making alkyd and polyester resins and fibers. Orthophthalic acid usually is not isolated because it loses a molecule of water so easily, forming phthalic anhydride, the commercially traded form of this strain of phthalic acid.

Exercises

1. What is the relationship of an acid to:
   a. an ester?
   b. an aldehyde?
   c. an acid anhydride?
   d. carbon dioxide?
2. How do you distinguish an acid from an alcohol?
3. What chemical property does acetic acid and phthalic acid have in common?

4. Adipic acid is to ____________ as cumene is to ____________ as ethylene dichloride is to ____________ as naphthalene is to ____________ as ethylbenzene is to ____________.

5. Insert in the middle column the intermediate compound structure, and identify in all three columns the names of the compounds:

Starting Compound	Intermediate Compound	Final Compound

CH_3-CH_2-OH

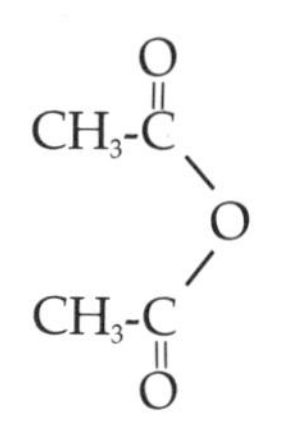

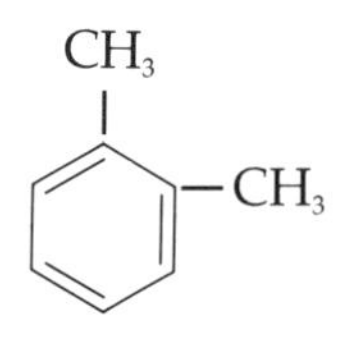

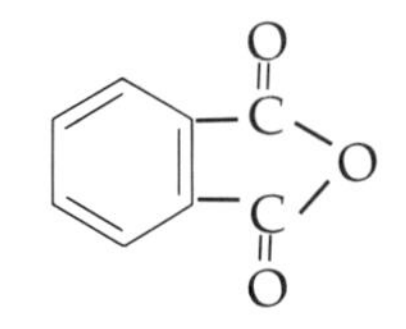

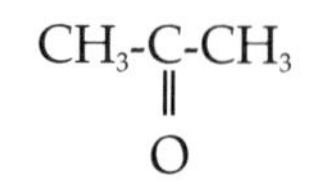

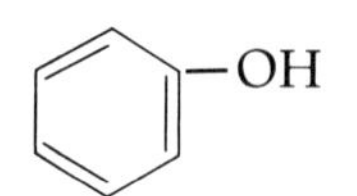

HOOC-$(CH_2)_4$-COOH

ACRYLONITRILE, ACRYLIC ACID, AND THE ACRYLATES

"I am a Bear
of Very Little Brain,
and long words Bother me."

Winnie-the-Pooh
A. A. Milne, 1882–1956

Acrylonitrile, acrylic acid and acrylates . . . they all sound alike; the molecules in Figure 17–1 look similar, and they are all covered in this same chapter. So they must be well connected. Right?

Well, not quite. First, they sound similar because at one time, one of them, acrylonitrile, was based solely on manufacture from acrolein, a pungent liquid whose roots in Latin are *acer,* meaning sharp, and *olere,* meaning smell. Acrylonitrile was made from acrolein, and acrylates were derivatives of acrylonitrile. But acrylates also are made from acrylic acid, which is also a derivative of acrylonitrile. So the name, acrylo, covers an extended family of relations.

Second, the similarity in structure has nothing to do with the relationship between the two. The double bonded group on the left of each molecule in Figure 17–1 is the vinyl group. What gives each of them their interesting and unique properties are the acid group ($-\overset{O}{\overset{\|}{C}}-OH$) in acrylic acid, the ester group ($-\overset{O}{\overset{\|}{C}}-OR$) in acrylates, and the nitrile or cyanide group (-CN) in acrylonitrile. Indeed, acrylonitrile used to be called vinyl cyanide, but that was before the petrochemical industry had good public relations people.

Third and finally, the three are covered in the same chapter because they are like the peaches, grapes and cherries in fruit salad. If any one of them are missing, you don't get quite the right flavor.

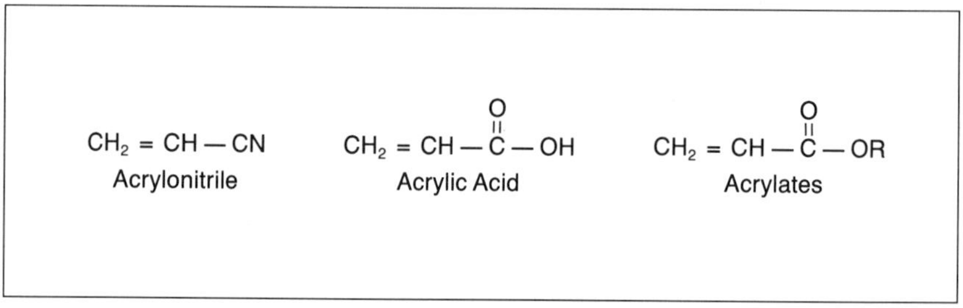

$$CH_2 = CH - CN$$
Acrylonitrile

$$CH_2 = CH - \overset{O}{\overset{\|}{C}} - OH$$
Acrylic Acid

$$CH_2 = CH - \overset{O}{\overset{\|}{C}} - OR$$
Acrylates

Fig. 17–1 Acrylonitrile, acrylic acid, and the acrylates

By now you should be wondering where this chapter will take you. So Figure 17–2 is the road map. You can see right away why the relationships are a little confusing. Acrylates can be made from propylene, acrylonitrile, or acrylic acid. Acrylic acid can be made from propylene or acrylonitrile. And acrylonitrile can be made from propylene, but it also is used to make other, unrelated but important things too, namely polymers or adiponitrile, the precurser to Nylon 66. If you keep this road map handy, the route through the next few pages will be easier.

Along the way there is one important processing fact to keep in mind. Acrylonitrile or acrylic acid may be an intermediate step to acrylates, but sometimes the intermediate is not isolated (separated or recovered) as a commercial product. That is what makes it difficult to separate discussion of the three chemicals in a neat, orderly way.

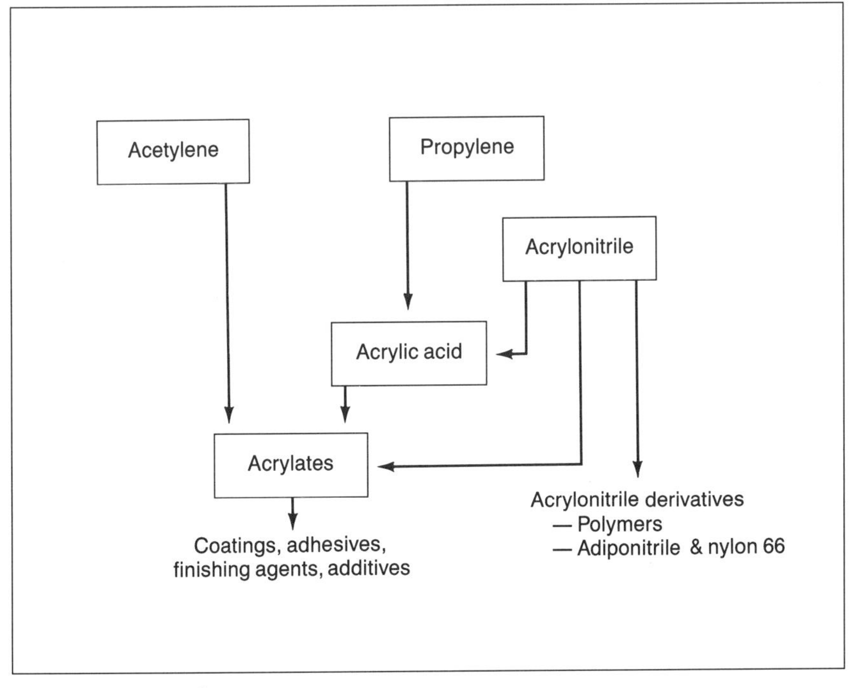

Fig. 17–2 A road map

Acrylonitrile

The nitriles are a group of compounds that can be thought of as derivatives of hydrogen cyanide, HCN. The hydrogen is removed and replaced by an organic grouping. In the case of acrylonitrile, the replacement is the vinyl grouping, $CH_2{=}CH$-, the same one encountered in styrene and vinyl chloride.

The original route to acrylonitrile was the catalytic reaction of HCN with acetylene. That was a combination of two compounds that together had all the characteristics you'd like to avoid—poisonous, explosive, corrosive, and on and on. But during World War II, acrylonitrile became very important as a comonomer for synthetic rubber (nitrile rubber). Later, the growth for acrylonitrile came from synthetic fibers like Orlon, Acrylon, and Dynel.

In the 1960s, like almost all acetylene technology, the HCN/C_2H_2 route to acrylonitrile gave way to ammoxidation of propylene. That word, ammoxidation, looks suspiciously like the contraction of two more familiar terms, ammonia and oxidation, and it is. Standard of Ohio developed a one-step vapor phase catalytic reaction of propylene with ammonia and air to give acrylonitrile.

$$CH_2=CH\text{-}CH_3 + NH_3 + 3/2\,O_2 \longrightarrow CH_2=CH\text{-}CN + 3H_2O$$

As a by-product, HCN is also formed, but there generally is a ready market for it. In fact, this process has become an important commercial source of HCN.

The Plant

The early ammoxidation plants were a two-step design. Propylene was catalytically oxidized to acrolein ($CH_2=CHCHO$). The acrolein was then reacted with ammonia and air at high temperature to give acrylonitrile. Most of this hardware has been replaced by the one-step process.

The Sohio technology is based on a catalyst of bismuth and molybdenum oxides. Subsequent catalyst improvements came from the use of bismuth phosphomolybdate on a silica gel, and more recently, antimony-uranium oxides. Each change in catalyst was motivated by a higher conversion rate per pass to acrylonitrile.

The propylene stream shown in Figure 17–3 can be either refinery grade (50–70% propylene) or chemical grade (90–95%). The propylene, ammonia, and oxygen are fed in a ratio of 1 to 1 to 2 to the vessel containing the catalyst. The vessel is called a fluidized bed reactor because the catalyst moves about like a fluid. The catalyst is usually a very fine, hard powder that flows very easily. As the reactants pass through the vessel, they are mixed with the catalyst. Because the catalyst particles are so small and there are so many, the total surface area of the catalyst exposed to the gaseous or liquid reactants is huge. So the yields from fluidized bed reactors are generally higher than fixed bed reactors. The main disadvantage is the loss of catalyst because of the difficulty of mechanically separating the particles out after the reaction is complete.

The ammoxidation reaction is carried out at about 800°F and 30 psi. Because it is highly exothermic, heat is removed continuously from the reactor by heat exchangers. The residence time of the reactants is

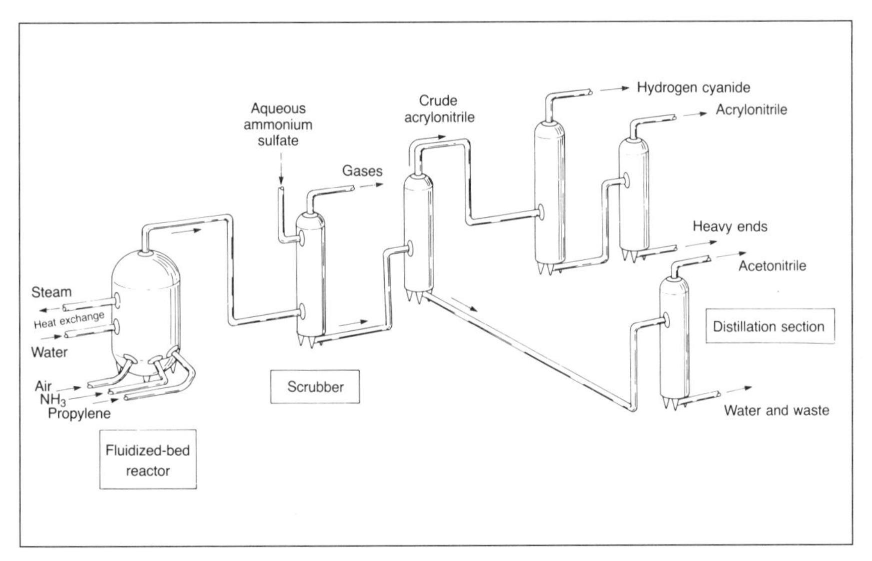

Fig. 17–3 Acrylonitrile plant

about three seconds. The effluent is quickly cooled and sent to be separated. The unreacted ammonia has to be "scrubbed" out by passing the effluent through a slightly acidic aqueous solution of ammonium sulfate. The gases (unreacted propylene, CO_2, and air) go overhead. The rest of the products go with the aqueous solution to be separated in a series of fractionating columns. The major byproducts are water, CO_2, acetonitrile, and hydrogen cyanide.

The major difficulties with these processes are controlling heat removal from the reactor; the stability of the catalyst, both mechanical and chemical; and catalyst loss. The latter two problems are due to the use of the fluidized bed reactor. Yields of acrylonitrile from this process are about 70%, based on propylene feed.

Material Balance

Feed:	
Propylene	1175 lbs.
Ammonia	475 lbs.
Air (excess O_2)	1631 lbs. O_2
Catalyst	small
Product:	
Acrylonitrile	1000 lbs.
Unreacted O_2 + NH_3 approx.	819 lbs.
HCN	100 lbs.
Water	1008 lbs.
Byproducts	354 lbs.

Other Processes

Catalysts also have been developed that will convert propane to acrylonitrile. Propane is almost always cheaper than propylene. The temperatures are higher (about 950°F) and the residence times a lot longer (about 15 seconds), so existing hardware generally cannot handle the new technology. The yields, based on propane, are also about 70%, but that hasn't been enough of an incentive to build new plants to shut down the existing ones.

Commercial Aspects

Uses. Acrylic fibers account for about half the acrylonitrile production. Orlon, Acrylon, and Dynel are polymers and copolymers of acrylonitrile. These fibers find extensive usage in apparel and household furnishings as well as in the industrial markets.

Nitrile rubber has declined in importance, but has been replaced by styrene-acrylonitrile (SAN) copolymers and acrylonitrile-butadiene-styrene (ABS) terpolymers. These plastics are relatively inexpensive, tough, and durable and were the first so-called engineering plastics to capture sizable pipe and auto parts markets.

A more recent use of acrylonitrile is its use to make adiponitrile, which is the feedstock used in Nylon 66 production. Acrylonitrile also has been found to be good treatment for cotton, making it resistant to mildew, heat, and abrasion, and more receptive to dyes.

Properties and Handling. Acrylonitrile is a colorless, flammable liquid with a boiling point of 171°F. It is traded commercially as technical grade (99%) and is bulk shipped in lined tank cars or trucks with the hazardous material markings. The linings are necessary due to the corrosive nature of acrylonitrile.

Acrylonitrile Properties

Freeze point	–117.0°F (–83.0°C)
Boiling point	171.1°F (77.3°C)
Specific gravity	0.811 (lighter than water)
Weight per gallon	6.7 lbs./gal.

Methacrylonitrile

Methacrylonitrile can be produced in the acrylonitrile plants by ammoxidation of isobutylene. This slightly different

molecule, $CH_2{=}C(CH_3){-}CN$, is copolymerized with acrylic acid, styrene, maleic anhydride, or isoprene to produce a wide variety of plastics and coatings.

ACRYLIC ACIDS

Many petrochemicals have been harnessed because they have two common characteristics: they're simple and they're reactive. Acrylic acid (AA) is the simplest organic acid that contains a double bond. It's that vinyl group again, $CH_2{=}CH$-, the same one found in acrylonitrile, styrene, and vinyl chloride. Because it's an acid and because it has the double bond, it's highly reactive. It readily undergoes polymerization (reacts with itself because of the double bond) and esterfication (reacts with alcohol because it's an acid).

The use of acrylic acid can be traced at least as far back as about 1900. It was an additive for paints and lacquers. Due to the tendency for acrylic acid to polymerize at low temperatures, it accelerated the "drying" process. The users probably didn't understand the chemistry of polymerization at the time, only that it worked.

Early routes to AA were complex and expensive. In 1927 the ethylene chlorohydrin process was introduced, but it was also still expensive, and not much commercial interest was stimulated in AA. In 1940 a process came literally right off the farm—pyrolysis of lactic acid, a waste product of the dairy industry found in sour milk. This route improved the economics of AA some, because of the availability of zero-cost raw material, the lactic acid. But the operating costs were still too high for rapid commercialization. It wasn't until the 1950s, with the Reppe process route to acrylic acid, starting with acetylene, and then in the 1960s, starting with propylene, did acrylics begin to take off.

With that meandering, historical background, it's better to switch to the subject of acrylates to deal with the specific manufacturing routes. In some, the acrylic acid, though formed in the process, never gets recovered as a commercial product. It just forms then converts to an acrylate.

Acrylic Acid Properties

Freezing point	53.8°F (12.1°C)
Boiling point	285.6°F (140.9°C)
Specific gravity	1.052 (heavier than water)
Weight per gallon	8.84 lbs./gal.

ACRYLATES (AND METHACRYLATES)

These are fun to read about because they end up in all sorts of products you're familiar with, but probably never thought too much about. To begin with, you need to understand what acrylates are.

If you take an alcohol (a compound with an -OH signature) and react it with an organic acid (one with a -COOH signature), the product is an ester (the -COOR signature) and the process is called esterfication. If the organic acid you use is acrylic acid, the ester is called an acrylate. And if the alcohol is, say, methyl alcohol, then the product is methyl acrylate, but not methacrylate. If you start out with methacrylic acid, then you get a methacrylate. And finally, if you use methyl alcohol and methacrylic acid, you get methyl methacrylate, which is a big star in petrochemicals.

You'll recall (of course) from Chapter I that the letter "R" is used as a substitute for a carbon group, like methyl, ethyl, etc. The general equation for esterfication of acrylic acid is:

$$\underset{\text{acrylic acid}}{CH_2=CH\text{-}\overset{\overset{\displaystyle O}{\|}}{C}\text{-}OH} + \underset{\text{alcohol}}{ROH} \longrightarrow \underset{\text{acrylate}}{CH_2=CH\text{-}\overset{\overset{\displaystyle O}{\|}}{C}\text{-}OR} + \underset{\text{water}}{H_2O}$$

Specifically, for the reaction with methyl alcohol,

$$CH_2=CH\text{-}\overset{\overset{\displaystyle O}{\|}}{C}\text{-}OH + CH_3OH \longrightarrow \underset{\text{methyl acrylate}}{CH_2=CH\text{-}\overset{\overset{\displaystyle O}{\|}}{C}\text{-}OCH_3} + H_2O$$

The major commercial acrylates are formed from the alcohols that should by now be familiar to you—methanol, ethanol, butanol, isobutanol, and 2-ethyl hexanol. The corresponding acrylates are shown in Table 17–1.

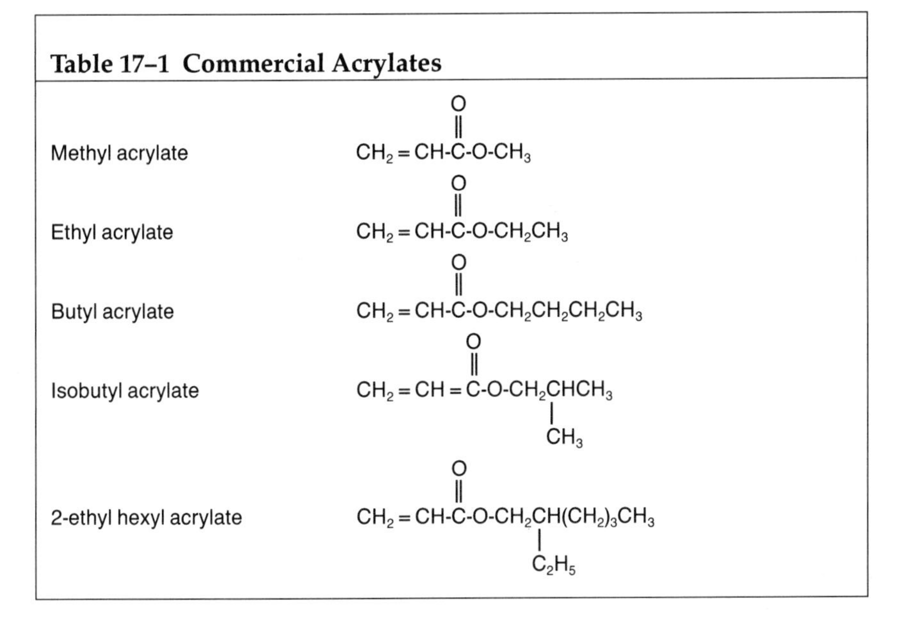

Table 17–1 Commercial Acrylates

Methyl acrylate	$CH_2=CH\text{-}\overset{O}{\overset{\|\|}{C}}\text{-}O\text{-}CH_3$
Ethyl acrylate	$CH_2=CH\text{-}\overset{O}{\overset{\|\|}{C}}\text{-}O\text{-}CH_2CH_3$
Butyl acrylate	$CH_2=CH\text{-}\overset{O}{\overset{\|\|}{C}}\text{-}O\text{-}CH_2CH_2CH_2CH_3$
Isobutyl acrylate	$CH_2=CH=\overset{O}{\overset{\|\|}{C}}\text{-}O\text{-}CH_2\underset{CH_3}{\underset{\|}{C}}HCH_3$
2-ethyl hexyl acrylate	$CH_2=CH\text{-}\overset{O}{\overset{\|\|}{C}}\text{-}O\text{-}CH_2\underset{C_2H_5}{\underset{\|}{C}}H(CH_2)_3CH_3$

Acetylene to Ethyl Acrylate

The Reppe process was commercialized in the 1950s. It involves the reaction of acetylene, carbon monoxide, and an alcohol (methyl, ethyl, etc.) to give an acrylic ester (an acrylate). The process is carried out at 125°F and 15 to 30 psi in a nickel carbonyl/aqueous hydrochloric acid solution. The nickel carbonyl acts as both a catalyst and a secondary source of carbon monoxide.

$$\underset{\text{Acetylene}}{HC\equiv CH} + \underset{\text{Carbon monoxide}}{CO} + \underset{\text{Ethanol}}{CH_3CH_2OH} \longrightarrow \underset{\text{Ethyl Acrylate}}{CH_2=CH\text{-}\overset{O}{\overset{\|\|}{C}}\text{-}OCH_2CH_3}$$

AA also can be made from this method by leaving out the alcohol and modifying the operating conditions. The conventional esterfication reaction to produce the acrylates can then be run. The lower molecular weight acrylates (methyl and ethyl) are usually produced via the "direct" technology. The higher molecular weight acrylates are usually made from methyl or ethyl acrylate by what chemists call trans-esterfication reaction. The higher weight alcohol does a little square dance with the acrylate, changing partners by replacing the methyl or ethyl group with a higher weight group such as a butyl or 2-ethyl hexyl group.

Reppe process yields are about 80%, but the usual acetylene drawbacks are present: hazardous materials handling and higher cost raw materials. As a result the acetylene route plants are not being duplicated as they wear out, giving way to the newer technologies.

Hydrolysis of Acrylonitrile

The cost of producing acrylonitrile dropped when the ammoxidation process was introduced in the 1960s. Then it became economical at that time to produce methyl and ethyl esters of acrylic acid by hydrolyzing acrylonitrile in the presence of alcohol. The hydrolysis and esterfication take place at the same time, in the presence of sulfuric acid at about 225°F. Yields are about 98%.

$$\underset{\text{Acrylonitrile}}{CH_2{=}CH\text{-}CN} + 2H_2O + H_2SO_4 \longrightarrow \underset{\text{Acrylic Acid}}{CH_2{=}CH\text{-}COOH} + \underset{\text{Ammonium Bisulfate}}{(NH_4)HSO_4}$$

$$\underset{\text{Acrylic Acid}}{CH_2{=}CH\text{-}COOH} + \underset{\text{Methanol}}{CH_3OH} \longrightarrow \underset{\text{Methyl Acrylate}}{CH_2{=}CH\text{-}\overset{\overset{\displaystyle O}{\|}}{C}\text{-}OCH_3} + H_2O$$

The process consumes the sulfuric acid as an ammonium bisulfate waste product, so it is expensive. So when propylene oxidation technology was developed, it became the preferred route.

Material Balance	
Hydrolysis of Acrylonitrile:	
Feed:	
Acrylonitrile	541 lbs.
Ethyl alcohol	469 lbs.
Water	184 lbs.
Sulfuric acid	980 lbs.
Product:	
Ethyl acrylate	1000 lbs.
Ammonium bisulfate	1150 lbs.
By-products	24 lbs.
Esterfication of Acrylic Acid:	
Feed:	
Acrylic acid	846 lbs.
Methyl alcohol	376 lbs.
Sulfuric acid catalyst	trace
Product:	
Methyl acrylate	1000 lbs.
Water	211 lbs.
By-products	11 lbs.

Catalytic Oxidation of Propylene

The newest and most commercially successful process introduced in the late 1960s involves vapor phase oxidation of propylene to AA followed by esterfication to the acrylate of your choice. Chemical grade propylene (90–95% purity) is mixed with steam and oxygen and reacted at 650–700 °F and 60–70 psi over a molybdate-tellurium metal oxide catalyst to give acrolein (CH_2=CH-CHO), an intermediate oxidation product on the way to AA. Acrolein is immediately passed through a second oxidation reactor to form acrylic acid. The reaction takes place at 475–575 °F, over a tin-antimony oxide catalyst. There are a few by-products, formic acid (HCOOH), acetic acid (CH_3COOH), low molecular weight polymers, carbon monoxide and dioxide. But overall yields of propylene to acrylic acid are high—85 to 90%.

Material Balance	
Propylene Oxidation	
Feed:	
Propylene	642 lbs.
Oxygen	735 lbs.
Catalyst	small
Product:	
Acrylic acid	1000 lbs
Water	250 lbs.
By-products	127 lbs.

Oxidative Carbonylation of Ethylene

A route not yet commercialized is the reaction of ethylene, carbon monoxide, and air to give AA. The ethylene is dissolved in acetic acid. The process takes place at 270°F and 1100 psi in the presence of palladium chloride-copper chloride catalyst. Yields are 80–85%. If the by-product and corrosion problems can be licked, the process will probably catch on.

Commercial Aspects

Uses. The most commercially important acrylates are ethyl-, butyl-, 2 ethyl hexyl-, and methyl-acrylate, in that order. Major markets include surface coatings, adhesives, textile finishing agent, paper coating, and additive. An important feature of the acrylates is that they readily polymerize if exposed to heat, light, oxygen, or peroxides. Most important, they polymerize in water to form a latex, which is a dispersion of solid particles in water, such as latex paints. A little diversion here might give a better understanding of the value of acrylates.

> Emulsion polymerization was developed as part of the synthetic rubber program during World War II. Take an acrylate monomer (an unpolymerized acrylate) and add it to water. It's immiscible—doesn't mix. If you add an emulsifying agent like soap (yes, soap), the acrylate becomes dispersable (miscible) in water. Now add a little water soluble catalyst. That induces polymerization. The acrylate monomer links itself chemically to other acrylate monomers, and as the polymer molecules grow to the right weight and size, they can be stabilized. The resulting mixture is called a latex. Add color pigment and you've got the basics for a latex paint.

Acrylic latices (more than one latex) find many uses in the field of coatings. Every amateur house painter appreciates the handling advantages:

1. When exposed to air and light, the latex will further polymerize to a hard coating at a moderate speed. (It "dries" fast.)
2. Before it polymerizes ("sets up"), it is soluble in water. (Easy clean-up of brushes and painter.)
3. After it polymerizes, it is stable and resists oxidation. (It's weather resistant and color-fast.)

4. During the drying process, only water vaporizes. With oil-based paints, naphtha or mineral spirits vaporize during drying. (Latex paints don't pollute the atmosphere.)

Handling. Acrylates are traded as technical grade (99% purity), inhibited or uninhibited. Usually they are sold with trace amounts of hydroquinone as an inhibitor.

Methyl and ethyl acrylates are toxic enough to require a hazardous shipping label, but butyl-, isobutyl-, and 2-ethyl hexyl-acrylates have high enough flash points to be considered safe.

An Example of Properties
Ethyl Acrylate

Freezing point	98.0°F (–72.0°C)
Boiling point	211.3°F (99.6°C)
Specific gravity	0.923 (lighter than water)
Weight per gallon	7.68 lbs./gal.

METHACRYLATES

The methacrylates are first cousins to the acrylates, but only one member of this branch of the family ever made it into commercial big time, methyl methacrylate (MMA). The most important feature of MMA is that it polymerizes into a transparent or translucent plastic.

$$CH_2=C(CH_3)-C(=O)-OCH_3$$

Methyl Methacrylate

Process

The production of MMA has long been accomplished by the old standby acetone cyanohydrin route. (See Fig. 17–4.) Acetone reacts with hydrogen cyanide in the presence of an aqueous solution of sodium hydroxide at 100–150°F to give acetone cyanohydrin. The MMA is then produced by hydrolyzing acetone cyanohydrin in the presence of 98% sulfuric acid and methyl alcohol. The two-step reaction occurs at about 200°F. After purification, overall yield is 80 to 85%.

Material Balance

Feed:	
Acetone	581 lbs.
Hydrogen cyanide	270 lbs.
Methanol	320 lbs.
Sulfuric acid (98%)	981 lbs.
Product:	
Methyl methacrylate	1000 lbs.
Ammonium bisulfate	1152 lbs.

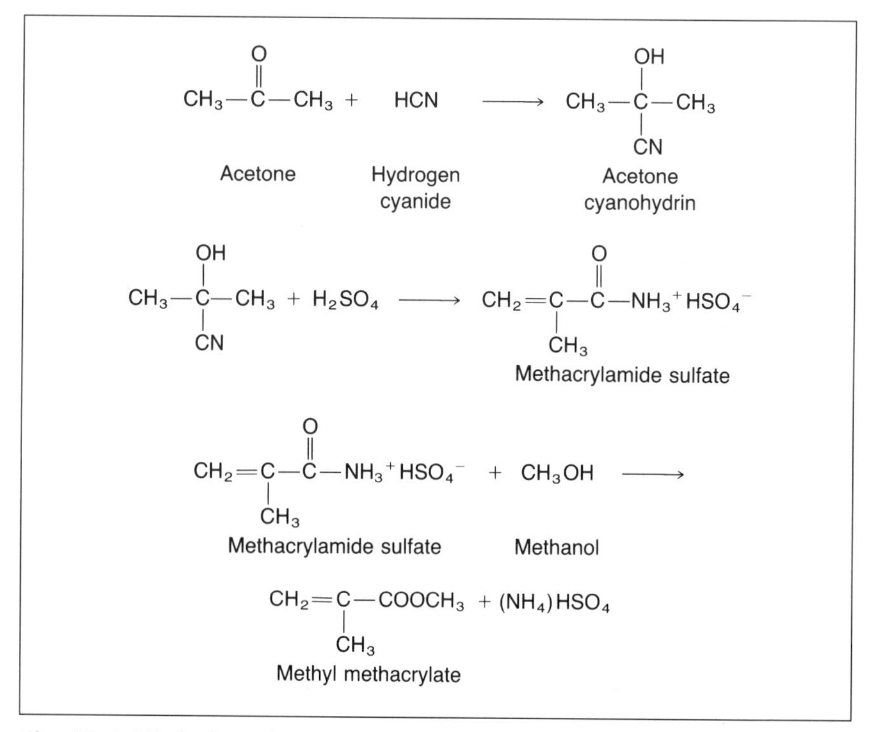

Fig. 17–4 Methyl methacrylate synthesis

Several newly developed alternate routes to MMA are gaining favor outside the U.S. Although these routes are more efficient and economical, American producers have thus far been loyal to the acetone cyanohydrin route. The plants are fully amortized and by staying with the old technology, producers can avoid the large capital investments associated with a new plant. In Europe and Asia, one route is based on ethylene and carbon monoxide. Another is tertiary butyl alcohol. The latter shows the most promise. It involves a catalytic two-step oxidation of TBA to methacrylic acid, which is then esterfied with methanol to MMA.

The BASF route to MMA involves a three-step process starting with ethylene. Reaction with carbon monoxide and hydrogen gives propionaldehyde (CH_3CH_2CHO) which becomes methacrolein ($CH_2{=}C(CH_3)\text{-}CHO$) by reaction with formaldehyde.

Mild oxidation with air yields methacrylic acid. Subsequent reaction with methanol gives MMA.

Commercial Aspects

Uses. The sole commercial use of MMA is polymers in various forms—cast sheets, latices, and molding and extrusion polymers. MMA polymers are best known for their use in the form of clear, transparent sheets with trade names like Plexiglass and Lucite. Applications include advertising signs, aircraft windows, desk tops, lighting fixtures, building panels, and plumbing and bathroom fixtures.

MMA is also used extensively as a copolymer with acrylates in latex paints and as a homopolymer in lacquers, since it's transparent.

MMA molding and extrusion polymers are used in the automotive industry for control dials, knobs, instrument covers, directional light covers, and tailgate lenses. The last two are probably the largest application of MMA molding powders.

MMA is also used in conjunction with other plastics to achieve translucent or transparent qualities. Transparent bottles, made by copolymerization of MMA with vinyl chloride, are gradually replacing glass containers. MMA is used in many of the same applications as acrylic latices and is also used as a comonomer with acrylonitrile to make acrylic fibers.

Methyl Methacrylate Properties

Freezing point	–54.8°F (–48.2°C)
Boiling point	212.2°F (100.1°C)
Specific gravity	0.938 (lighter than water)
Weight per gallon	7.86 lbs./gal.

Properties and Handling. MMA is a colorless, sweet smelling, volatile liquid that boils at 212°F. MMA readily polymerizes with itself, and usually has trace amounts of hydroquinone added as an inhibitor. MMA is traded as technical grade and is shipped in lined tank cars, tank trucks, and drums. The hazardous material warnings are required on all shipments.

Chapter XVII in a nutshell...

Acrylonitrile, C_2H_3CN or $CH_2=CH\text{-}CN$, has the characteristic nitrile signature group, -CN. The double bond between the carbons makes "acrylo" useful in polymerizations as an intermediate in the manufacture of acrylates and adiponitrile for Nylon 66 production. The primary route to acrylo is the reaction of ammonia and oxygen with propylene. The poor match of atoms in and out results in only 70% yield.

Acrylic acid, $CH_2=CHCOOH$, has the characteristic signature group for acid. Acrylic acid can be made from propylene or from acrylonitrile, and is generally used to make acrylates.

The acrylates (for example, ethyl acrylate, $CH_2=CH\text{-}COOCH_2CH_3$) are esters of acrylic acid, so they end in the suffix *-ate* and have the characteristic signature -COOR.

The methacrylates, which are commercially even more important than the acrylates, are esters of methacrylic acid and are used extensively in coatings, plastics, and adhesives.

Exercises

1. What feedstock do acrylonitrile and acrylic acid have in common? How about acrylates and acrylic acid?
2. What are the major differences between a fixed bed catalyst process and a fluidized bed process? What are the advantages of a fluidized bed process? The disadvantages?
3. The vinyl grouping, $CH_2{=}CH$-, imparts chemical reactivity to an organic compound. List a half dozen or so compounds containing this reactive grouping.
4. Esterfication and dehydration reactions both produce water. How do they differ?
5. Would "olefins" be considered suitable feedstock for manufacturing acrylonitrile, propyl acrylate, MMA, and methacrylonitrile? Elaborate.

XVIII

MALEIC ANHYDRIDE

"How many apples
fell on Newton's head
before he took
the hint?!"

■

Robert Frost, 1874–1963

The unlikely molecule in Figure 18–1 is a cyclic anhydride known by several names: 2-butene-1, 4-dicarboxylic acid anhydride; cis-butene-dioic acid anhydride; maleic anhydride (MA); and when you've been in the business a long time, "maleic."

The name, maleic anhydride, came about in the same fashion as any number of compounds early in the petrochemical business. Many organic acids and their derivatives were given common names based on some early observations, their special source in nature, or on some special feature of their structure. MA was first isolated in the 1850–75 era by dehydration of malic acid, a sugar acid found in apple juice. The Latin word for apple is malum.

Hence, malum, malic, maleic. The suffix, anhydride, which follows each alias of MA, has a simple definition: a compound derived by the loss of a molecule of water from two carboxyl groups (-COOH).

You may wonder why a chemical with such an unusual structure is so popular. The answer, as always, is reactivity. But in this case, MA is thrice blessed. If you refer to Figure 18–1, there is reactivity associated with:

- the anhydride group on the right;
- the double bond on the left; and
- the carboxylic acid grouping that (re-)forms when MA is mixed with water. That's the group that gave up the water molecule in the previous paragraph that formed MA to begin with.

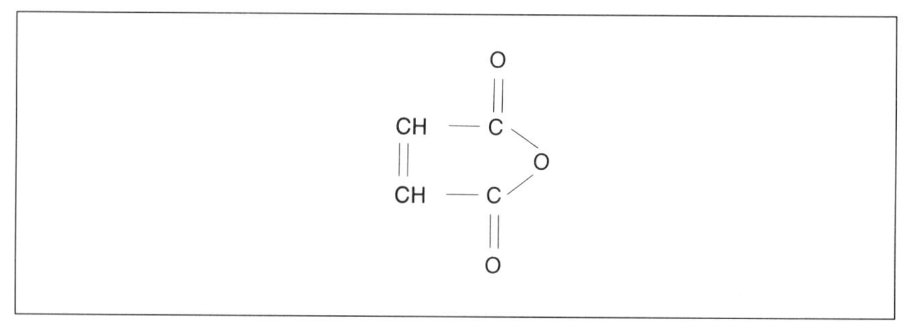

Fig. 18–1 Maleic anhydride

MA can be produced from several different feedstocks—benzene, normal butenes, and normal butane. The popularity of any one of them has swayed with the economic winds that set the feedstock prices. Benzene was the original choice in the 1940–50 period. Butene came on strong in the 1950s, and faded quickly as they became surplus in refineries, and then short again. The divergence between the prices for normal butane and butene or benzene has stimulated interest in butane. Moreover, normal butane has a yield advantage. If you study the MA molecule for a moment, you'll see that it has only four carbons. When benzene (C_6H_6) is the feedstock, two carbon atoms must be eliminated to form MA. They end up as a waste product, CO_2. When butane (C_4H_{10}) is the feedstock, the atoms eliminated are the lightweights, hydrogen. The theoretical yields of MA are 1.26 pounds of MA per pound of benzene but 1.75 pounds of MA per pound of normal butane. So the yield advantage is no small economic factor.

The benzene and butane routes are very similar. Benzene route hardware is often adaptable to the butane route because the pressures, temperatures, and even the catalyst are the same. For that reason, benzene plants have been converted to butane plants fairly cheaply. Today, practically all the MA produced in the U.S. is based on butane feed. Elsewhere in the world, the favorite feed is benzene.

Process

The key to the reactions in Figure 18–2 is the incredible ability the catalyst has to rearrange the atoms and their bonds. The catalyst for all three feedstocks is V_2O_5 (vanadium pentoxide) and a promoter. In the case of benzene, a MoO_3 (molybdenum trioxide) promoter is added. For butane and butylene, it's P_2O_5 (phosphorous pentoxide.) Consider the changes that take place with benzene in only a single pass through the oxidation reactor, a lapse time of one second:

a. The benzene ring is opened and two of the carbon atoms are cleaved off as CO_2.
b. The resulting butene molecule undergoes selective oxidation where the two terminal methyl groups are converted to carboxylic acid groups.
c. A molecule of water is then lost, giving rise to the heterocyclic anhydride grouping and thus, MA.
d. All of this occurs without oxidizing the reactive double bond.

If you set out to accomplish all that in a chemical process, you surely wouldn't expect to be lucky enough to find something as selective and powerful as V_2O_5. Clarke's Third Law is true: "Any sufficiently advanced technology is nearly indistinguishable from magic."

Fixed bed plants. In this type of plant, the process flow for all three feeds looks like the plant in Figure 18–3. The feed and compressed air are mixed, vaporized in a heater, and then charged to the reactor. The ratio of air to hydrocarbon is generally about 75 to 1. The reactor consists of a bundle of tubes packed with the catalyst. The feed temperature is 800–900°F, depending on the feed. The reaction time is extremely quick, so the feed is in contact with the catalyst for only 0.1 to 1.0 second.

Like most oxidation reactions, this one is exothermic, and extremely so. That's why the catalyst is in tubes—coolant is pumped past

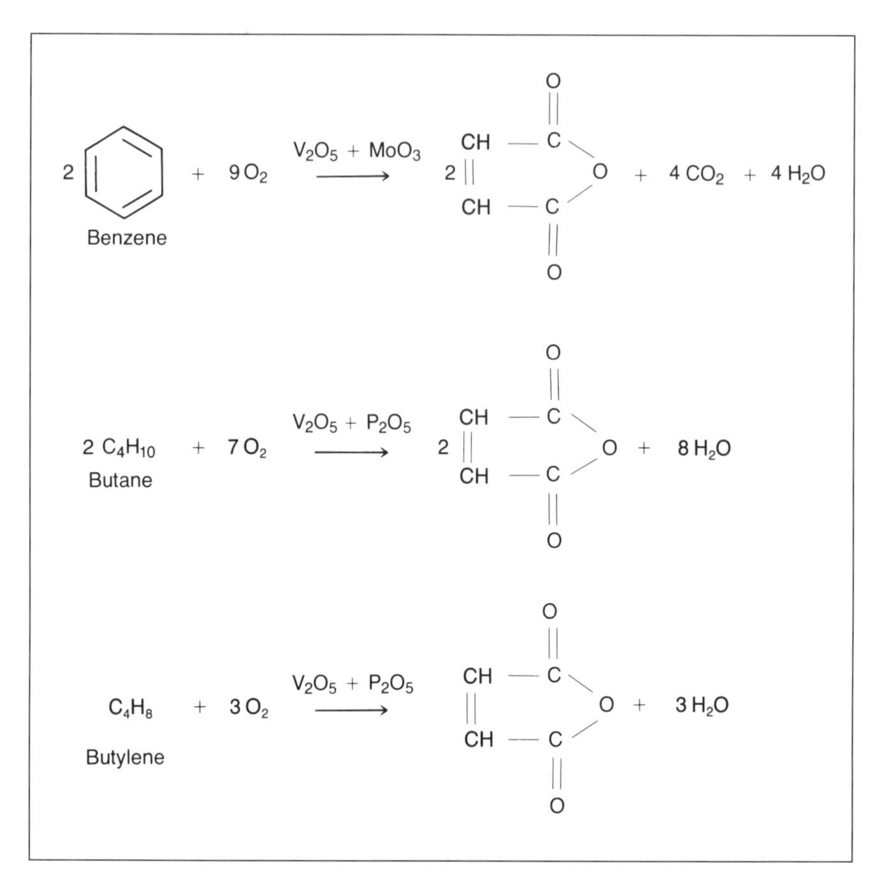

Fig. 18–2 Routes to maleic anhydride

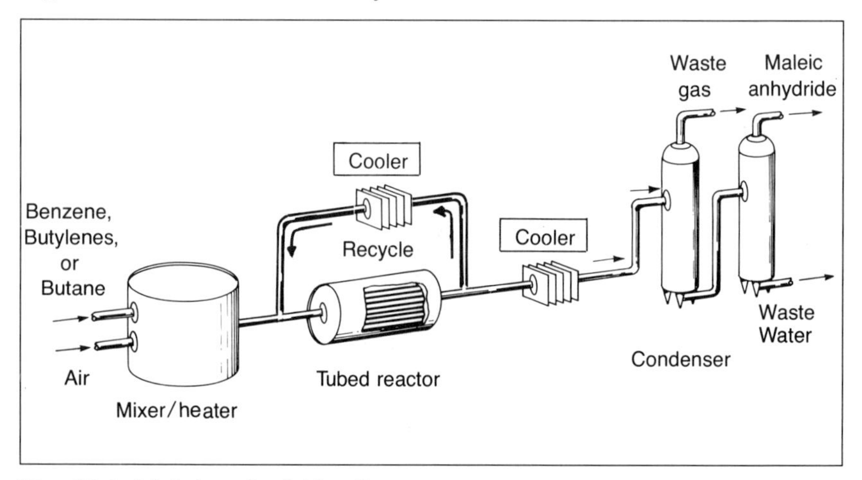

Fig. 18–3 Maleic anhydride plant

the tubes to keep the reaction temperature from running away. Also, the reaction is self-sustaining. That is, the reaction gives off enough heat to keep itself going. Once it gets hot enough to get started, it'll continue with no more heat added.

The effluent gas from the reactor contains about 50% maleic acid (not maleic anhydride). The balance is some unreacted feed, CO_2, water, and some miscellaneous waste products. A recycle stream is passed through a cooler and recharged to the reactor. The purpose is not only to take another pass at the feed but to dilute the feed with some already-made maleic acid. That helps to disperse the heat of reaction and to control the operating conditions.

The bulk of the effluent is run through coolers and separators to get a clean maleic acid stream which is easily dehydrated (removal of a water molecule) by applying heat in a distillation column. The dehydrated maleic acid is maleic anhydride.

Fluidized Bed Plants. New plant technology is emerging which differs from the scheme shown in Figure 18–3 in that the catalyst moves around with the feed during the reaction rather than staying fixed in the reactor tubes. The design, called fluidized bed technology, uses catalyst in a powdered form that is so mobile that it can be pumped like a liquid or blown like a vapor.

The catalyst is mixed with air where it picks up oxygen atoms, then is blown together with butane into a reactor where the chemical reaction takes place. The effluent from the reactor is a mixture of catalyst, MA, water vapor, and a little feedstock. The catalyst is removed by a contraption called a cyclone, which uses centrifugal force to spin the solid, heavier catalyst particles out of the mixture. The MA and feedstock (butane, etc.) are then separated for recovery (the MA) or recycle (the feedstock).

Commercial Aspects

Uses. About 60% of the MA produced is used to make unsaturated polyester and alkyd resins. Polyester resins are used in the fabrication of glass fiber reinforced parts. Applications include boat hulls, automobile body parts, patio furniture, shower stalls, and pipe. Alkyd resins are mostly used in coatings (paint, varnish, lacquers, and enamels). MA also is widely used as a chemical intermediate in the manufacture of plasticizers and dibasic acids (fumaric, maleic, and succinic). Several agricultural chemicals are based on maleic anhydride, the best known being Malathion.

Maleic Anhydride Properties	
Freezing point	127.4°F (53.0°C)
Boiling point	391.5°F (199.7°C)
Specific gravity	0.934
Physical appearance	white needles or flakes with an acrid odor

Properties, Grades, and Handling. MA melts at 127°F, so at normal temperatures it is a white solid with an acrid odor. The vapors are highly toxic, and will burn your eyes and give you a skin rash. It's soluble in water and many organic solvents.

MA is available commercially in 99% purity in both molten (liquid, above 127°F) and solid forms (flakes, pellets, rod, or briquets). MA is often shipped in fiber drums or bags. Heated tank cars or trucks are used for liquid shipments. Because of the toxic fumes, the hazardous materials designation must be posted on all shipments.

Chapter XVIII in a nutshell...

Maleic anhydride is a cyclic anhydride with one double bond in the ring and two double bonded oxygens hanging off the ring. The resulting reactivity leads to maleic's use in making polymers, unsaturated polyesters, alkyd resins, plasticizers, and dicarboxylic acids.

Maleic can be made by oxidation of butane or benzene. The process would otherwise be virtually impossible without the use of vanadium pentoxide as the catalyst. It enables extensive reconfiguration of either feedstocks molecular structure into the anhydride structure.

Exercises

1. ____________, ____________, and ____________ are three important anhydrides derived from what feedstocks? What are their structural differences and why is one of these anhydrides more reactive?
2. Clark's Third Law could apply to the highly selective V_2O_5 catalyst for manufacturing MA. Name two other catalysts that might fall under Clark's Third Law.
3. What are two good reasons for choosing between butylene, butane, or benzene as the feedstock for MA?

ALPHA OLEFINS

"I do not view
the process with
any misgivings."

■

Winston Churchill, Tribute to the Royal Air Force, House of Commons, August 20, 1940

Alpha olefins occupy a strangely unique niche in the petrochemicals industry. Their name and their chemical structure imply they are a basic building block. In a way they are. But they are also derivatives of ethylene, and they are grown almost in a way that polymers are grown, just not as long. One other fact that might be surprising is that one of the alpha olefins is butene-1, a petrochemical covered in Chapter VI, The C_4 Hydrocarbon Family. The reason the name alpha olefin didn't come up is that generally those C_4's result from cracking larger molecules. Although the alpha olefin, butene-1, and all the other alpha olefins once came from this route, now they come from just the inverse type of process—they're grown from the bottoms up.

Alpha olefins are straight chain hydrocarbons having a double bond in the number one carbon-carbon position. That's called the alpha position, and hence the name, alpha olefin. (There are beta, gamma, etc., compounds around, too.) The chains can have as few as four carbons (butene-1) or more than 30 (written $C_{30}+$). And they all have the double bond in the alpha position, as you can see in Figure 19–1. At one time, alpha olefins could have either an odd or even number of carbon atoms, but through a quirk of the manufacturing processes now in vogue, only even number carbon count alpha olefins are produced.

The interest in alpha olefins as a group lies in the reactivity of the double bond, just like the styrene, vinyl chloride, ethylene, propylene, or acrylonitrile. But individually, the alpha olefins of varying chain length have quite different physical characteristics and therefore different applications. For example, the C_4 alpha olefin is a gas at room temperatures, while the C_6 through C18's are liquids and the $C_{20}+$ are waxy solids.

CH_3-CH_2-$CH=CH_2$

butene-1

CH_3-CH_2-CH_2-CH_2-$CH=CH_2$

hexene-1

CH_3-CH_2-$(CH_2$-$CH_2)_n$-$CH=CH_2$ where n=2, 3, . . . to 13+

C_8 to $C_{30}+$ alpha olefins

Fig. 19–1 Alpha olefins

Historical Development

In the early 1960s, alpha olefins were produced by thermally cracking waxy paraffins found in crude oils. The process consisted in subjecting the wax to high enough temperatures to cause cleavage of the carbon to carbon bonds in the long wax chain molecules. Because of the absence of extra hydrogen, the cracking process leaves a double bond at the end of the resulting molecules. The various chain lengths were then separated by distillation. This route worked okay, but was ripe for improved efficiency. It was energy intensive so it was expensive. It also resulted in a high proportion of branched olefins (having side-chains) because of the characteristics of the feedstocks. If the feed had side-chains, the alpha olefins were likely to have side-chains.

In the late 1960s, the oligomerization route was introduced. (Oligomer comes from the Latin root olig-, meaning a few; and mer, meaning part, as in monomer, polymer, etc.) The process is based on "growing" chains by addition of ethylene molecules. That overcame the problem of branching and left the expensive, energy-intensive processing to the olefin producers. Within a few years, the oligomerization process had completely replaced all the older technology.

But growing oligomers with ethylene results in chains of only even numbers of carbons. Wax cracking gave both even and odd numbers. Wouldn't that be an obstacle to full commercialization of the oligomerization process? As any tourist guide in a third world country will tell you, "No problem!" As it turns out, the demand for alpha olefins is almost entirely for ranges of chain lengths, not specific carbon count chains. Examples are shown in Table 19–1. One exception is butene-1, which is used by itself in several applications, such as polybutylene manufacture and as a polyethylene comonomer.

By the late 1980s the alpha olefin market had grown to over two billion pounds. That was not just because the new technology had been developed. Simultaneously, a broad range of applications for all the alpha olefins expanded rapidly—surfactants, synthetic lubricants, plasticizer alcohols, fatty acids, mercaptans, comonomers, biocides, paper and textile sizing, oil fields chemicals, lube oil additives, plastic processing aids, and cosmetics. Quite a list.

Table 19–1 Alpha Olefin Applications

C_4-C_8	polymers and polyethylene comonomer
C_6-C_8	low molecular weight fatty acids and mercaptans
C_6-C_{10}	plasticizer alcohols
C_{10}-C_{12}	synthetic lubricants and additives, detergent amine oxides and amines
C_{14}-C_{16}	detergent alcohols and nonionics
C_{16}-C_{18}	lube oil additives and surfactants
C_{20}-C_{30} +	oil field chemicals and wax replacement

Manufacturing Alpha Olefins

Ethylene oligomerization is accomplished by successive addition of ethylene molecules.

$$\underset{\text{ethylene}}{CH_2{=}CH_2} + \underset{\text{ethylene}}{CH_2{=}CH_2} \longrightarrow \underset{\text{butene-1}}{CH_3\text{-}CH_2\text{-}CH{=}CH_2}$$

$$\underset{\text{butene-1}}{CH_3\text{-}CH_2\text{-}CH{=}CH_2} + \underset{\text{ethylene}}{CH_2{=}CH_2} \longrightarrow \underset{\text{hexene-1}}{CH_3\text{-}(CH_2)_3CH{=}CH_2}$$

$$\underset{\text{hexene-1}}{CH_3\text{-}(CH_2)_3\text{-}CH{=}CH_2} + \underset{\text{ethylene}}{CH_2{=}CH_2} \longrightarrow \text{etc., up to } C_{30}+$$

The catalyst that ignites this process was named after its inventor, Karl Ziegler, the notable German chemist. He found that triethyl aluminum could, under the right pressure and temperature conditions, be used as a kind of a root for growing hydrocarbon chains. Triethyl aluminum is a compound of aluminum with three ethyl groups attached. When subjected to high pressures and temperatures and an excess of ethylene, a hydrogen atom at the terminal end of the ethyl group can be displaced by ethylene, starting the growth of a chain. Other ethylene molecules will also continue adding at the end of the new chain, as long as there are sufficient ethylene molecules around and the temperature and pressure conditions are right.

When the process of chain growth is satisfactorily completed, separation of the three hydrocarbon chains that are connected to the aluminum atom is accomplished by a displacement reaction. The chain-laden aluminum compound (called trialkyl aluminum compounds) is subjected to still higher temperatures and pressure. This causes an ethylene molecule to displace the long linear carbon chain. As the separation is made, triethyl aluminum is reformed, making a recylable root for another go-around.

The displacement reaction:

$$\underset{\text{aluminum alkyl}}{Al[CH_2\text{-}CH_2\text{-}(CH_2\text{-}CH_2)_n\text{-}CH_2\text{-}CH_3]_3} + \underset{\text{ethylene}}{CH_2{=}CH_2} \longrightarrow$$

$$\underset{\text{triethyl aluminum}}{Al(CH_2\text{-}CH_3)_3} + \underset{\text{an alpha olefin}}{3\ CH_3\text{-}CH_2\text{-}(CH_2\text{-}CH_2)_n\text{-}CH{=}CH_2}$$

The above chemistry is sometimes accomplished simultaneously in one reactor and sometimes in two separate reactors. In the former the triethyl aluminum catalyst is lost; in the latter it is recycled. Sometimes the displacement compound is butene-1 or hexene-1, depending on the chain length of the final alpha olefin desired and the change in operating conditions necessary to effect the displacement reaction.

Typical yields of the two processes are shown in Table 19–2. Actually there is more flexibility than what the table suggests. In the early 1970s, most of the demand for alpha olefins was in the C_{10}-C_{16} range. That was at a time when environmental concerns were escalating over the foaming being caused by phosphate-based household detergents. As a consequence the demand for bio-degradable detergents that happened to be based on linear, even-numbered C_{10}-C_{16} range alpha olefins precursors took off. The demand skewed towards the middle of the range of alpha olefin production. But by continuous development of the process, the production of alpha olefins at either end of the distribution was mitigated. There is continuing progress in matching production to distribution, even as new applications of both ends of the range change.

Table 19–2 Alpha Olefin Production Distribution (percent by weight)

	One-step process	Two-step process
C-4	13–16	5–10
C-6 to C-10	50–40	42–63
C-12 to C-18	27–30	25–50
C-20 +	10–14	2–3

Other routes

More recently, other catalysts have been developed which challenge the effectiveness of the Ziegler catalyst. A nickel/phosphine catalyst is used in a two-step process. The displacement step in this process also has the flexibility to convert whatever lighter or heavier alpha olefins are created into detergent range olefins, C_{10}-C_{16}. These, however, have internal, rather than alpha double bonds. That is, they aren't alpha olefins, and they contain odd as well as even number carbon chains. They are useful, however, for making alcohols for the surfactant market. The attractiveness of this process is the manufacturer's choice of having all of the linear olefins (alpha and internal) in the detergent range, C_{10} to C_{16}.

The Process

The general flow and the reactions for a typical alpha olefin process are shown in Figure 19–2.

High purity ethylene gas plus recycle ethylene are fed to a compression chamber, compressed, and then fed along with catalyst previously dissolved in a suitable solvent into parallel horizontal reactors, each containing a single pipe coil immersed in water. Reaction conditions are 350–425°F and 2000–3000 psi. The reaction is exothermic and the temperature is controlled by a back pressure control valve on the steam being evolved from the reactor shell. Olefin conversion per pass can be as high as 60% with yields approaching 95%. (Conversion relates to how much ethylene disappears in one pass; yield relates to how much of it ends up in the finished product, not in the by-products. See the appendix for further discussion.)

The reactor effluent is cooled and fed to the ethylene separator for recovery of unreacted gaseous ethylene. The liquid phase is filtered to remove small amounts of polymer and then treated with aqueous caustic to remove the catalyst. The dissolved light ends (C_2 and C_4 olefins) are separated by suitable fractionating towers in series. A portion of the ethylene is purged to remove methane and ethane, and the remaining ethylene is recycled to the compressor. The butene-1 is removed to storage.

The C_6 and heavier olefins are then separated via a series of atmospheric and vacuum fractionation towers. Multiple towers or columns are required to separate the heavier olefins.

Properties and Handling Characteristics

Typical properties for the alpha olefins produced by ethylene oligomerization are given in Table 19–3. You can find in the table that as the carbon count increases, purity declines. The impurities are branched chains and internal olefins (beta, gamma, etc.) These variations have more opportunity to form as the molecules get longer—Murphy's Third Law in operation again.

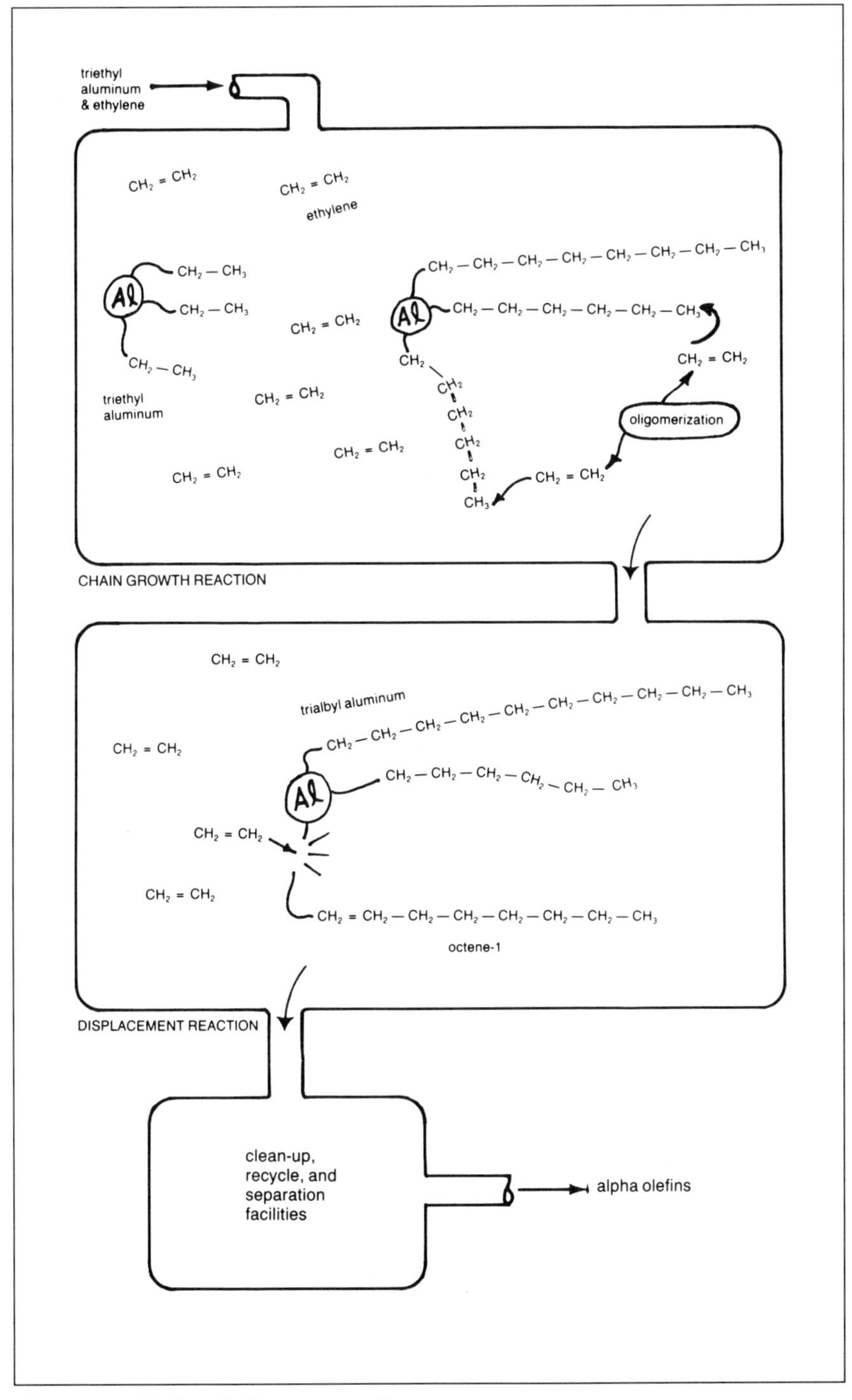

Fig. 19–2 Alpha olefin process flow

Table 19–3 Typical Alpha Olefin Properties

	Butene-1 C_4	Hexene-1 C_6	Octene-1 C_8	Decene-1 C_{10}	Do-decene-1 C_{12}	Tetra-decene-1 C_{14}	Hexa-decene-1 C_{16}	Octa-decene-1 C_{18}	C_{20}-C_{24}	C_{24}-C_{28}	C_{36} +
% Purity	99.0	97.5	96.5	94.5	94.5	93.5	92.0	91.0	ca. 88	ca. 86	ca. 84
Distillation Range — °F	43.0	144-147	248-257	334-347	401-428	454-491	518-572	—	—	—	—
°C	6.1	62-64	120-185	168-175	205-220	240-255	270-300	—	—	—	—
Color and Appearance	colorless, clear							white, waxy solid			
Freeze Point °F	—	—	—	—	—	—	40	65	110	145	180
Flash Point °F	—	less than 60		120	180	225	270	290	375	380	510
Specific Gravity	0.595	0.678	0.718	0.745	0.763	0.776	0.785	0.792	0.799	0.819	0.830
Weight per Gallon, lbs.	5.00	5.70	6.04	6.26	6.41	6.52	6.59	6.65	6.71	6.88	6.95

The flash points shown are standard measures of flammability. They are the temperature at which the liquid gives off enough flammable vapor to the surrounding air to ignite. The lower the flash point, the more flammable and dangerous the compound. The products are shipped in tank trucks, tank cars and barges which may or may not have heating coils. The higher carbon number alpha olefins, C_{18}+, will require heating coils. Nitrogen is required for the liquid C_6 to C_{16}'s because exposure to air gradually causes a reaction with oxygen, producing peroxides. Stabilizers also are used to minimize peroxide build-up.

DOT classifies butene-1 as a flammable compressed gas, hexene and octene as flammable liquids, and decene-1 as a combustible liquid. All four must carry required shipping placards. The remaining alpha olefins are not regulated by the DOT.

Commercial Aspects

About 80–90% of the alpha olefins produced in the United States are used as comonomers in LLDPE (linear low density polyethylene) and HDPE (high density polyethylene), plasticizer alcohols, polyalpha olefins for use in synthetic lubricants, detergent alcohols, surfactants, and polybutylene. The remaining 10–20 percent are used in the other applications listed in Table 19–1.

Chapter XIX in a nutshell...

Alpha olefins are straight chain olefins that have a double bond in the number one (alpha) carbon-carbon position. Because they are now made by linking ethylene molecules together, alpha olefins have only even number carbon counts. Alpha olefins with 4, 6, 8, to 30 or more are commercially available.

Alpha olefins are made by oligomerization, growing them on an aluminum root by adding ethylene until the desired size is reached. The chain length is determined by the pressure and temperature conditions in the reactor.

The variety of alpha olefin application is extensive, including polymers, surfactants, synthetic lubricants, lube oil additives, plasticizer alcohols, mercaptans, and fatty acids.

Exercises

1. What is the difference between the process for producing alpha olefins and the one described in Chapter XII for producing higher alcohols?
2. Why are there only even number carbon count alpha olefins any more?
3. Why do you think there is a distribution of different carbon count alpha olefins rather than just one?

COMMENTARY THREE

"It is the beginning
of the end."

■

Charles Maurice de Talleyrand, 1754–1838, commenting on the Battle of Bordino in 1812.

REVIEW

There are many areas of commonalty in the hodgepodge of petrochemicals just covered. Synthesis gas, the underground petrochemical building block, is the route to some alcohols including methyl, isobutyl, and normal butyl alcohol. Yet hardly anyone knows anything about synthesis gas because it's not a traded commodity. But you can't get much more basic a building block than carbon monoxide and hydrogen.

Ethyl alcohol is made by direct, catalyzed hydration. Isopropyl alcohol is a product of indirect hydration, which occurs at much lower pressures and temperatures than the direct method and is therefore more economical.

The process to make 2-ethyl hexanol starts with propylene and synthesis gas, but it takes dimerization and hydrogenation to form the proper carbon chain and alcohol group, -OH.

The processes for the ketones, acids, acrylonitrile and the acrylates, and maleic anhydride defy simple summarization. Just read them individually. The chemical structures for most of them are shown in the following table. It might help if you try to recognize the signature group in each molecule.

As one of the newest petrochemicals to make it big, alpha olefins have found two niches. One is in the petrochemicals industry as precursors for detergents and specialty chemicals. The other is in this book, appropriately sandwiched in between the usual petrochemical derivatives and the polymers. They are derivatives, but the process for creating them is more like polymerization than any other derivative process.

FOREWORD

The best is yet to come. The most interesting chapters in this book are the next few. One reason is that the polymers are so much a part of your everyday life. Much of what you touch and see nowadays (and even what you are) is made in whole or in part from polymers. So in the next few chapters you'll be coming across words that you were familiar with before you ever got into the petrochemical business.

The other reason is that polymer chemistry pulls together a lot of what you've already learned about petrochemicals. So many of them end up in the polymerization processes.

The polymer chapters tend to be long. There's a lot to cover under each topic. As a matter of fact, before you get to read about *the* polymers in Chapters XXI and XXII, you need to read *about* polymers in Chapter XX. It's a big body of chemistry and chemical engineering, but these chapters should give you a handle on it.

Chemical Structures of Various Compounds

Name	Nickname	Chemical configuration
Synthesis gas	Syngas	H-H and CO
		ALCOHOLS
Methyl alcohol	Wood alcohol	$CH_3\text{-}OH$
Ethyl alcohol	EA	$CH_3\text{-}CH_2\text{-}OH$
Isopropyl alcohol	IPA	$CH_3\text{-}\underset{\underset{OH}{\vert}}{CH}\text{-}CH_3$
Normal butyl alcohol	NBA	$CH_3\text{-}CH_2\text{-}CH_2\text{-}CH_2\text{-}OH$
2-ethyl hexanol	2-EH	$CH_3\text{-}CH_2\text{-}CH_2\text{-}CH_2\text{-}\underset{\underset{CH_2\text{-}CH_3}{\vert}}{CH}\text{-}CH_2\text{-}OH$
		KETONES
Acetone	DMK	$CH_3\text{-}\underset{\underset{O}{\Vert}}{C}\text{-}CH_3$
Methyl ethyl ketone	MEK	$CH_3\text{-}\underset{\underset{O}{\Vert}}{C}\text{-}CH_2\text{-}CH_3$
Methyl isobutyl ketone	MIBK	$CH_3\text{-}\underset{\underset{O}{\Vert}}{C}\text{-}CH_2\text{-}\underset{\underset{CH_3}{\vert}}{CH}\text{-}CH_3$
		ACIDS
Acetic acid	—	$CH_3\text{-}\underset{\underset{O}{\Vert}}{C}\text{-}OH$
Acrylic acid	—	$CH_2=CH\text{-}\underset{\underset{O}{\Vert}}{C}\text{-}OH$
Adipic acid	—	$HO\text{-}\underset{\underset{O}{\Vert}}{C}\text{-}(CH_2)_4\text{-}\underset{\underset{O}{\Vert}}{C}\text{-}OH$
Terephthalic acid	TPA, PTA	$HO\text{-}\underset{\underset{O}{\Vert}}{C}\text{-}C_6H_4\text{-}\underset{\underset{O}{\Vert}}{C}\text{-}OH$

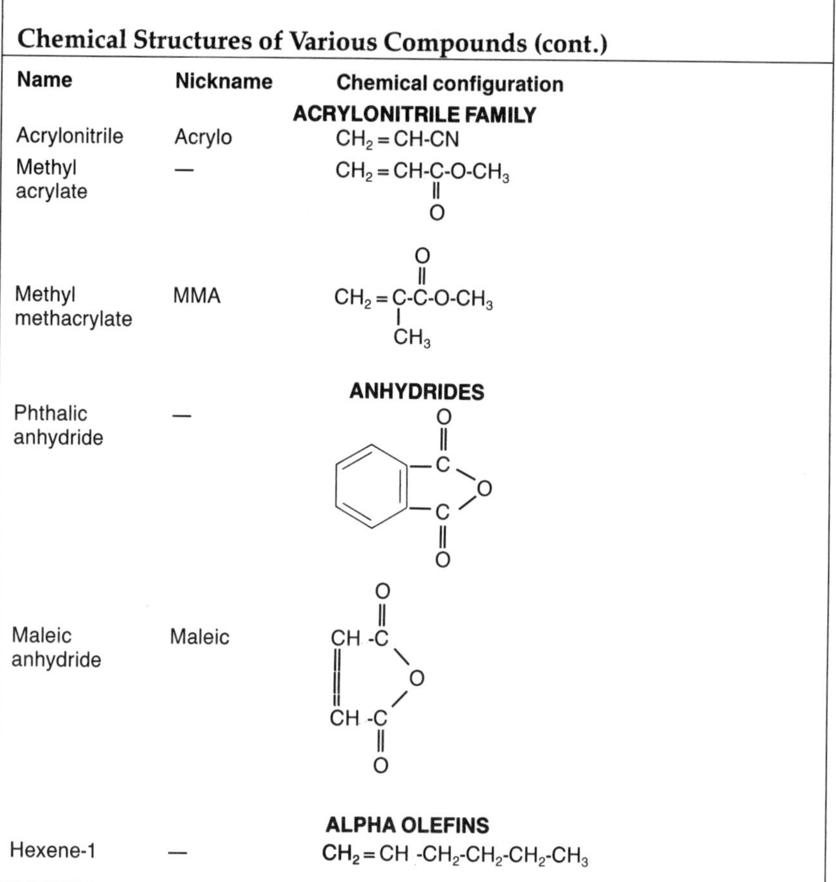

Chemical Structures of Various Compounds (cont.)

Name	Nickname	Chemical configuration
		ACRYLONITRILE FAMILY
Acrylonitrile	Acrylo	$CH_2=CH\text{-}CN$
Methyl acrylate	—	$CH_2=CH\text{-}C(=O)\text{-}O\text{-}CH_3$
Methyl methacrylate	MMA	$CH_2=C(CH_3)\text{-}C(=O)\text{-}O\text{-}CH_3$
		ANHYDRIDES
Phthalic anhydride	—	
Maleic anhydride	Maleic	
		ALPHA OLEFINS
Hexene-1	—	$CH_2=CH\text{-}CH_2\text{-}CH_2\text{-}CH_2\text{-}CH_3$

THE NATURE OF POLYMERS

"These are ties which,
though light as air,
are links of iron."

■

On Conciliation With America
Edmund Burke, 1729–1797

Polymers are a pretty complicated subject. That's why they're treated in three successive chapters. In this one you'll find a number of ways people classify polymers. It's quite an inventory:

Resins vs. Plastics

Thermoplastics vs. Thermosets

Homopolymers vs. Copolymers

Bifunctional vs. Polyfunctional

Linear vs. Branched vs. Cross-Linked

Addition vs. Condensation

The problem is that polymer chemistry became a virtual explosion of ideas and options as it developed in the 1950s and was further commercialized in the 1960s. There's no easy way to cover polymers other than to wade through. But go ahead. It's not hard and you'll learn a lot.

A Little History

The first partially synthetic polymer dates back to 1869, when cellulose (wood pulp) was nitrated (nitrocellulose). The cellulose became processible, and with the further addition of camphor, which acted as a plasticizer, it became a clear, tough, moldable product with the trade name "Celluloid." It was widely used at the end of the nineteenth century in the form of combs, brushes, photographic film, and shirt collars.

Not much commercial development took place until the chemistry of polymerization was starting to be understood in the 1930s and 1940s. Commercialization of some of the key polymers happened as follows:

1869–Nitrocellulose
1908–Bakelite (first synthetic commercial plastic)
1919–Polyvinyl acetate
1931–The polyacrylates
1936–Polyvinyl chloride
1938–Nylon and polystyrene
1942–Polyethylene and polyesters
1947–Epoxies
1953–Polyurethanes
1957–Polypropylene
1964–Polyimides
1973–Polybutylenes
1977–Linear low density polyethylene

More than 50% of the chemical industry in the U.S. is now based on or dependent on polymers.

Classifying Polymers

For a field of scientific and engineering endeavor, polymers have one of the more sloppy sets of nomenclature. Ask six people in the business to give you definitions of resins and plastics and you'll get at least six different answers. Almost everyone will tell you that they're both polymers, and that's right. Some will tell you they're interchangeable. Strictly speaking, they're wrong.

A lot of people will tell you that plastics will flow when they're heated or reheated, but that resins are set permanently so that heating them won't do anything. A lot more people, particularly in the fabrication end of the business, think resins are unfabricated polymers, and plastics are resins after they've been molded and/or set by the process, extruders, etc.

If you trace the word resin back far enough, you'll find that it was originally defined as a low molecular weight, natural polymer which is an exudate of (it exudes from) vegetable or non-vegetable matter. Examples are rosin (from pine trees), shellac (from insects), and both frankincense and myrrh (aromatic gums from an East African and an Asian species of tree). Resins like these do not flow if heat and pressure are applied, like plastics do. They decompose or melt. (This definition of resin is obsolete in commerce today.)

So you'll get no neat definition of resins and plastics here. But you'll know to be careful when someone else uses either term. Now, as for polymer, that's defined as a high molecular weight molecule formed by joining, in a repetitive pattern, one or more types of smaller molecules.

Polymers fall into one of two major classes, thermoplastics and thermosets. Despite the fact that thermosets have been around much longer, thermoplastics make up about 80% of the industry output. Thermoplastics are linear polymers that can be re-softened a number of times, usually by applying heat and pressure. They can be dissolved in solvents (suitable for that purpose). That's not true for thermosets once they're set. After they're formed or cured (by heat and/or pressure), these cross-linked three-dimensional polymers become non-melting and insoluble. Thermosets actually decompose under heat before they melt.

Both thermoplastics and thermosets can be used in four of the five major application areas: plastics, elastomers, coatings, and adhesives. But, only thermoplastics can be used in making fibers. During the spinning and drawing process of fiber processing, it's necessary to orient the molecules. Only unbranched, linear polymers (not thermosets) are capable of orientation.

Polymers result from polymerization—the chemical combination of a large number of molecules of a certain type, called monomers.

Monomers can be bifunctional (capable of joining up with two other monomers) and tri- or polyfunctional (each may join up with three or more monomers). When bifunctional monomers react with each other, you get linear thermoplastic polymers. If tri- or polyfunctional monomers react, you get cross-linked polymers, most of which are thermosets. Figure 20–1 illustrates these variations.

In some cases, the monomers react with themselves to form homopolymers:

ethylene to polyethylene
vinyl chloride to polyvinyl chloride
styrene to polystyrene

In other cases, and actually most of the time, two or more different monomers react to form copolymers:

butadiene and styrene to Buna S Rubber
styrene and acrylonitrile to SAN
ethylene glycol and terephthalic acid to
polyethylene terephthalate (PET)

Some of the copolymer variations are also illustrated in Figure 20–1.

Making Polymers

The polymerization process can be an addition reaction or a condensation reaction. Addition involves monomers containing a carbon-carbon double bond like this: $CH_2=CH$-R. If R is hydrogen, then the monomer is ethylene. If R is chlorine, then the monomer is vinyl chloride; if it's a methyl group, then the monomer is propylene; a benzene ring, then it's styrene; and so on to more complicated structures.

Condensation polymers generally result from simple reactions involving two different monomers, each containing different functional groups. The usual example is terephthalic acid and ethylene glycol to make polyester. The two monomers will react in such a way that a small molecule like water or methanol is given off as a by-product.

Addition Polymerization

This type of polymerization is a technique for adding monomers end to end. It involves three steps: initiation, propagation, and termination.

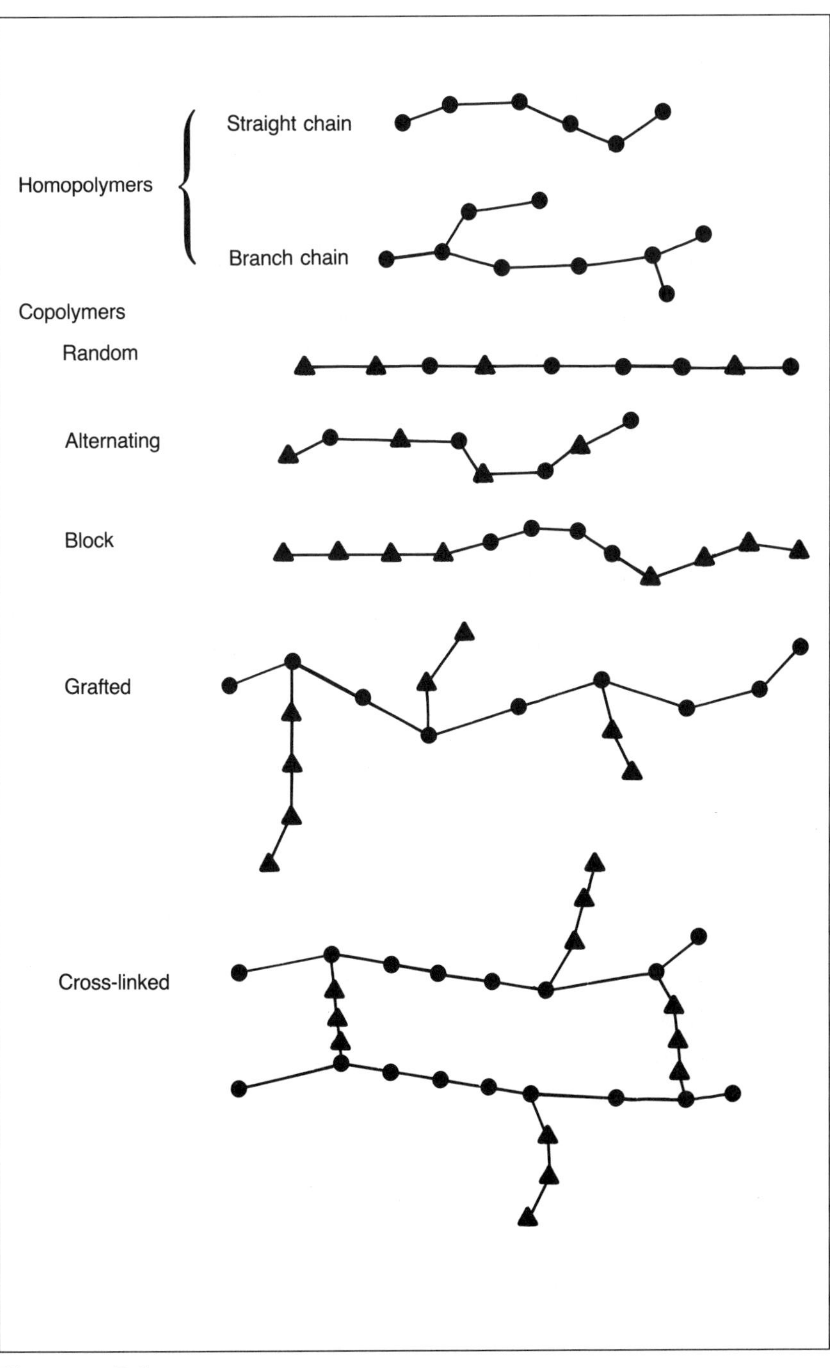

Fig. 20–1 Polymer structures

Initiation. The trick here is to get the reaction started. Usually a catalyst is used. A typical catalyst is some kind of organic peroxide, molecules that are somewhat unstable. When they're heated, they decompose and turn into highly reactive free radicals. As you'll recall, a radical is an almost-complete molecule, but all the valence requirements are not satisfied. So it is very anxious to meet up with some other molecule to satisfy its valence needs. The free radical, in the presence of an abundance of monomers, say a million to one ratio, will react with a monomer molecule. It becomes part of the molecule. In doing so, the unsatisfied valence condition now transfers to the end of the monomer. A new radical is formed. That's the start of the initiation step.

Propagation. Now the new radical collides and reacts with another monomer molecule to give a new larger radical, which in turn reacts with another monomer, and so on, and so on. This chain growth continues until propagation is terminated. The propagation or growth step in a commercial process usually takes a couple of seconds. The number of monomers in the chain is at least a 1000 or more. By employing special catalysts some polyethylenes are produced with up to 150,000 repeating units, and all in a few seconds.

Termination. A number of mechanisms are used to stop the propagation or growth step. A common way occurs when the monomer concentration is so low that the free radical chains dimerize. That is, they collide with each other and form a stable polymer, with all valence requirements satisfied.

Branch polymers or branch chains are short or long chains which are at right angles to the original chain's backbone (Figure 20–1). Short chains can be deliberately added using comonomers such as butene-1 or hexene-1. Long chain branching often happens in high pressure polyethylene processes. In the propagation step, a "growing" polymer radical extracts an "inside" hydrogen atom from a "finished" polymer chain. That now becomes a new polymer radical (at that site) and a chain can start growing there. Sometimes this new reaction is facilitated by a chain transfer agent. Iso-butane, propylene, and dodecyl mercaptan do well.

Cross-linked polymers occur when polymer chains are linked together at one or more points (other than their ends). Cross-linking can occur when the monomers involved are polyfunctional. That is, they have more than two active sites where links can be attached. So they grow like long chains, but also they link up with each other. Cross-linking also can be initiated by adding special agents. (Like Charles Goodyear did when he accidentally spilled some sulfur into a vessel of molten natural rubber. In the process, he "discovered" vulcanization, cross-linking with sulfur atoms.) Cross-linked polymers lose their moldability, even when they're reheated, because the molecules are chemically bonded in place, and do not slip and slide.

The length, branching, and cross-linking of the polymers are controlled by the timing of the three steps. A lot of initiating catalyst will result in an abundance of free radicals. When that happens, the concentration of the monomer goes down rapidly as a relatively high number of polymers start growing all at once. This results in early termination and a large number of small (low molecular weight) polymers. The properties of these polymers would be very different (maybe better or worse) than the converse, a relatively small number of large (high molecular weight) polymers. Usually the large molecules are what you're after.

Copolymers. Mixtures of two or more different bifunctional monomers can undergo addition polymerization to form copolymers. Why copolymerize? Well, polymers have different properties that depend on their composition, molecular weight, branching, crystallinity, etc. Many copolymers have been developed to combine the best features of each monomer. For example, polystyrene is low cost and clear, but it is also brittle with no toughness. It needs internal plasticization. By copolymerizing styrene with small amounts of acrylonitrile or butadiene, the impact and toughness properties are dramatically improved.

Another reason for copolymerization is to insert functional grouping in the polymer. A functional group is one that is easily reacted. For example, copolymerization of styrene with acrylonitrile, $CH_2{=}CH{-}CN$, involves only the double bond, leaving the newly formed copolymer with the active functional group -CN, available for subsequent reaction. The copolymer might be reacted later with itself or another monomer to give a cross-linked thermoset.

A third reason for interest in copolymers is crystallinity. Transparency and translucency are greatly affected by crystalline properties, which can be regulated by copolymerization.

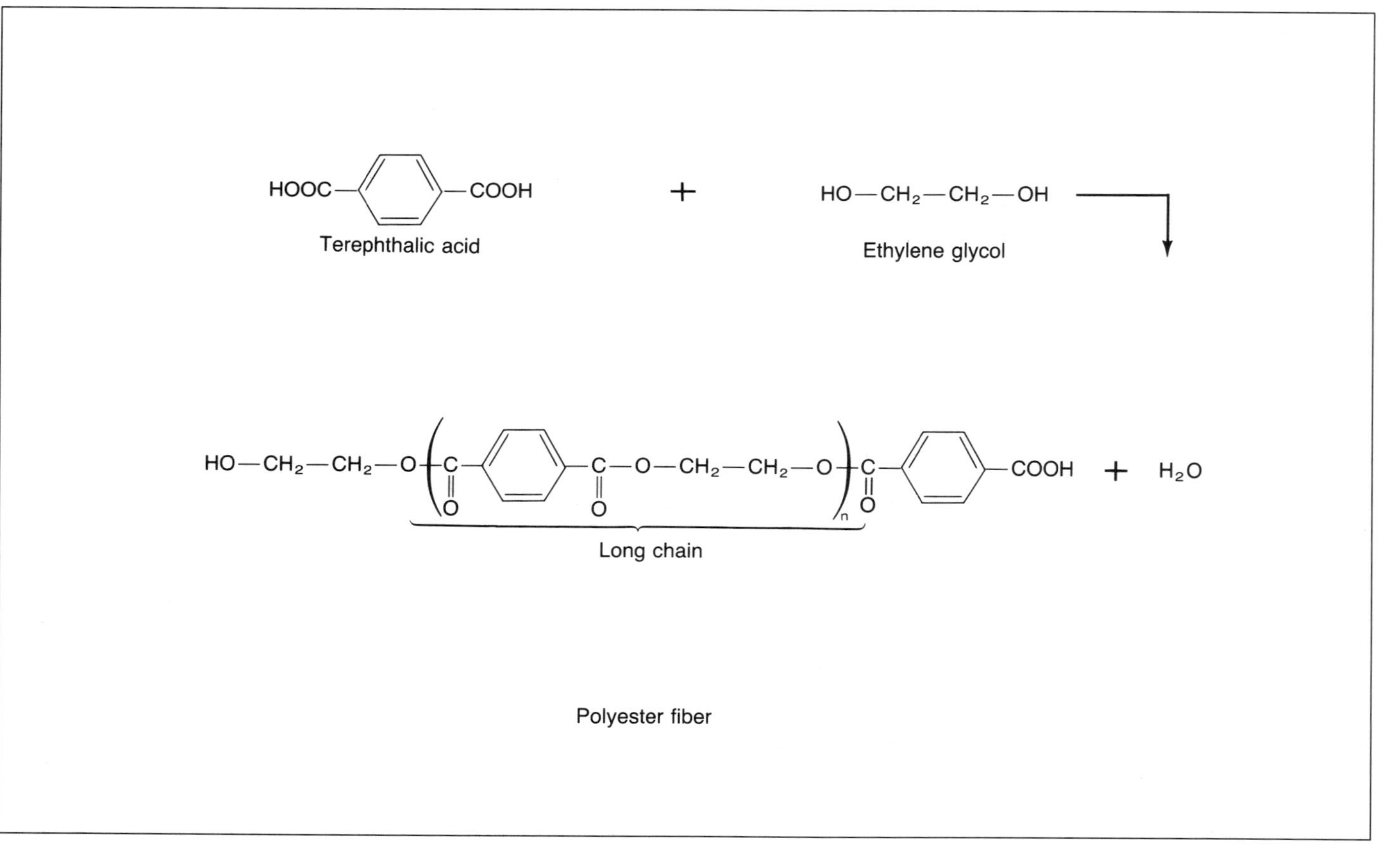

Fig. 20–2 A condensation polymerization

Condensation Polymerization

Condensation polymers are always copolymers. They are always formed by a series of chemical reactions involving two reactive sites in each which can join to form linking bonds. By-products are usually given off and are generally small molecules such as water, methanol, or hydrogen chloride.

Because two reactive sites are necessary, bifunctional monomers are often used in condensation polymerization. A bifunctional monomer includes molecules with two identical signature groups in them. Examples would be terephthalic acid and ethylene glycol, both shown in Figure 20–2. When a bifunctional monomer like either is used, the polymerization step is end-to-end, forming long chains. The reaction in Figure 20–2 is a "simple" esterfication of ethylene glycol with terephthalic acid to make polyethylene terephthalate, which is polyester fiber or Dacron.

There are numerous bifunctional monomers used in condensation polymerization. Some of the more popular signature groups that turn up frequently are shown in Figure 20–3. Important copolymers made by condensation include epoxies, nylon, polyesters, polycarbonate, and polyimides. As always, there are exceptions and one is Nylon 6 made by a ring opening reaction of caprolactam. All of these will be covered in the next two chapters.

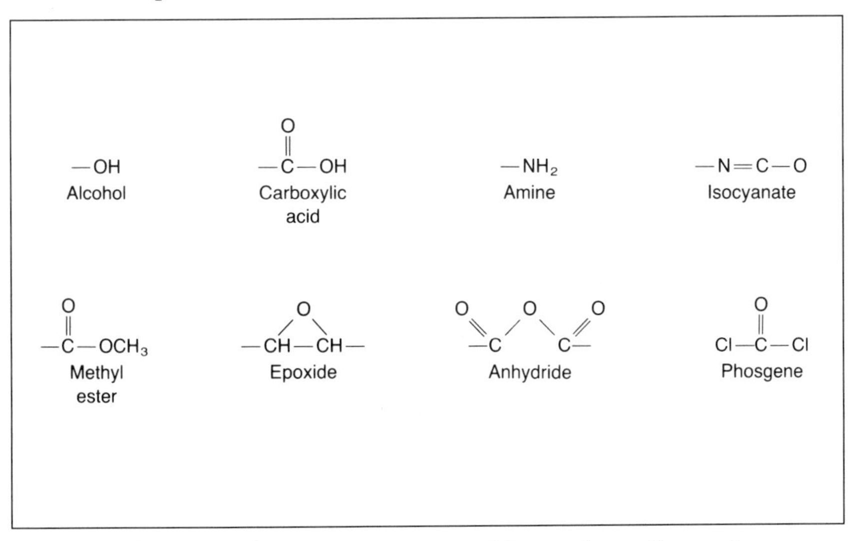

Fig. 20–3 Common signature groups used in condensation polymers

Thermosets

Thermosets cure into non-melting, insoluble polymers. Frequently, the curing needs heat, pressure, or catalyst to proceed. Often the final cure, which is nothing more than completion of the cross-linking, takes place in the fabrication or molding operations. The chemistry is about the same as you saw in the thermoplastics, but there are more reactive sites per monomer. (They are polyfunctional.) Consequently, more three-dimensional cross-linking takes place.

The simplest way to achieve three-dimensional cross-linking is to use monomers with three or more reactive sites. Examples are maleic anhydride, butadiene, isoprene, epichlorohydrin, pyromellitic dianhydride, and trimethylol propane.

As an illustration, consider some of the elastomers. In its natural state, rubber lacks toughness. In the 1939 accident already mentioned, Goodyear found that by reacting latex rubber (natural or synthetic) with sulfur, he could improve its strength and toughness, and increase its temperature properties. What he was doing was cross-linking rubber with sulfur in a process now commonly called vulcanization. The reaction with polyisoprene rubber is shown in Figure 20–4. Other synthetic rubbers such as butyl rubber (from iso-butylene and butadiene), Buna S (butadiene and styrene), Buna N (butadiene and acrylonitrile), and neoprene (chloroprene) can all be vulcanized to thermoset rubbers. They all have the polyfunctional configuration that makes cross-linking by sulfur possible. After vulcanization, they are tough, resist deforming, and are heat and cold insensitive, resistant to solvents, and non-conductive.

Other important thermosets include phenolics like Bakelite, epoxy resins, polyimides, and polyurethanes.

Methods of Polymerization

Most polymers are made by one of four processes commercially available. Each process has advantages related to the monomer being used and the end use of the polymer.

Bulk Polymerization. This is the simplest method. Monomers and initiator are mixed in the reactor shown in Figure 20–5 and heated to the right temperature. The bulk process is suitable for condensation polymers because the heat of reaction is low (it gives off less heat.)

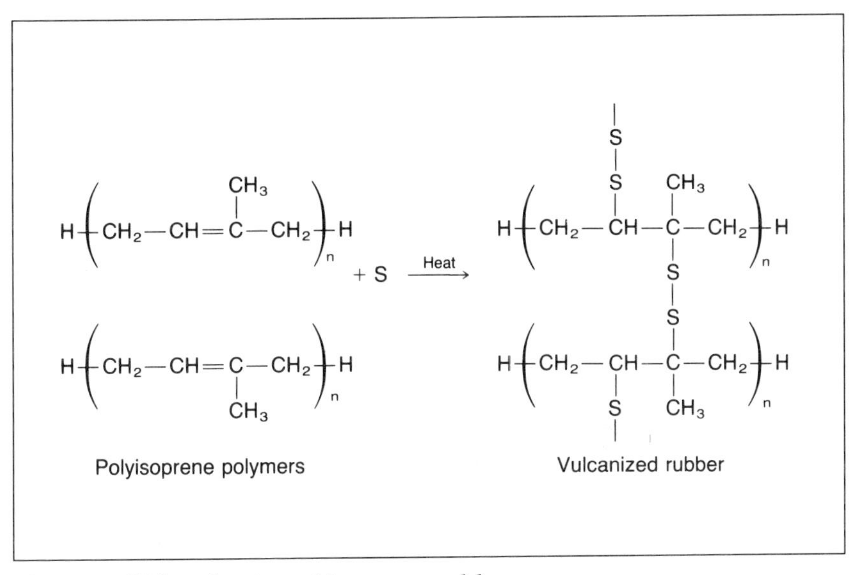

Fig. 20–4 Vulcanization of isoprene rubber

Polymerization of methyl methacrylate to Plexiglass is done in the bulk process. High pressure polymerization of ethylene is done this way also. But other addition polymerizations frequently become too exothermic and without adequate heat removal system, the reaction tends to "run away" from optimum conditions.

Solution Polymerization. Highly exothermic reactions can be handled by this process. The reaction is carried out in an excess of solvent which absorbs and disperses the heat of reaction. The excess solvent also prevents the formation of slush or sludge, which sometimes happens in the bulk process when the polymer volume overtakes the monomer. The solution process is particularly useful when the polymer is to be used in the solvent, say like a coating. Some of the snags with this process: it's difficult to remove residual traces of solvent, if that's necessary; the same is true of catalyst if any is used. This process is used in one version of a low pressure process for high density polyethylene and for polypropylene.

Suspension Polymerization. In this process, monomers and initiator are suspended as droplets in water or a similar medium. The droplets are maintained in suspension by agitation (active mixing). Sometimes a water soluble polymer like methylcellulose or finely divided clay are added to help stabilize or maintain the droplets. After formation, the polymer is separated and dried. This route is used commercially for vinyl-type polymers such as polyvinyl chloride and polystyrene.

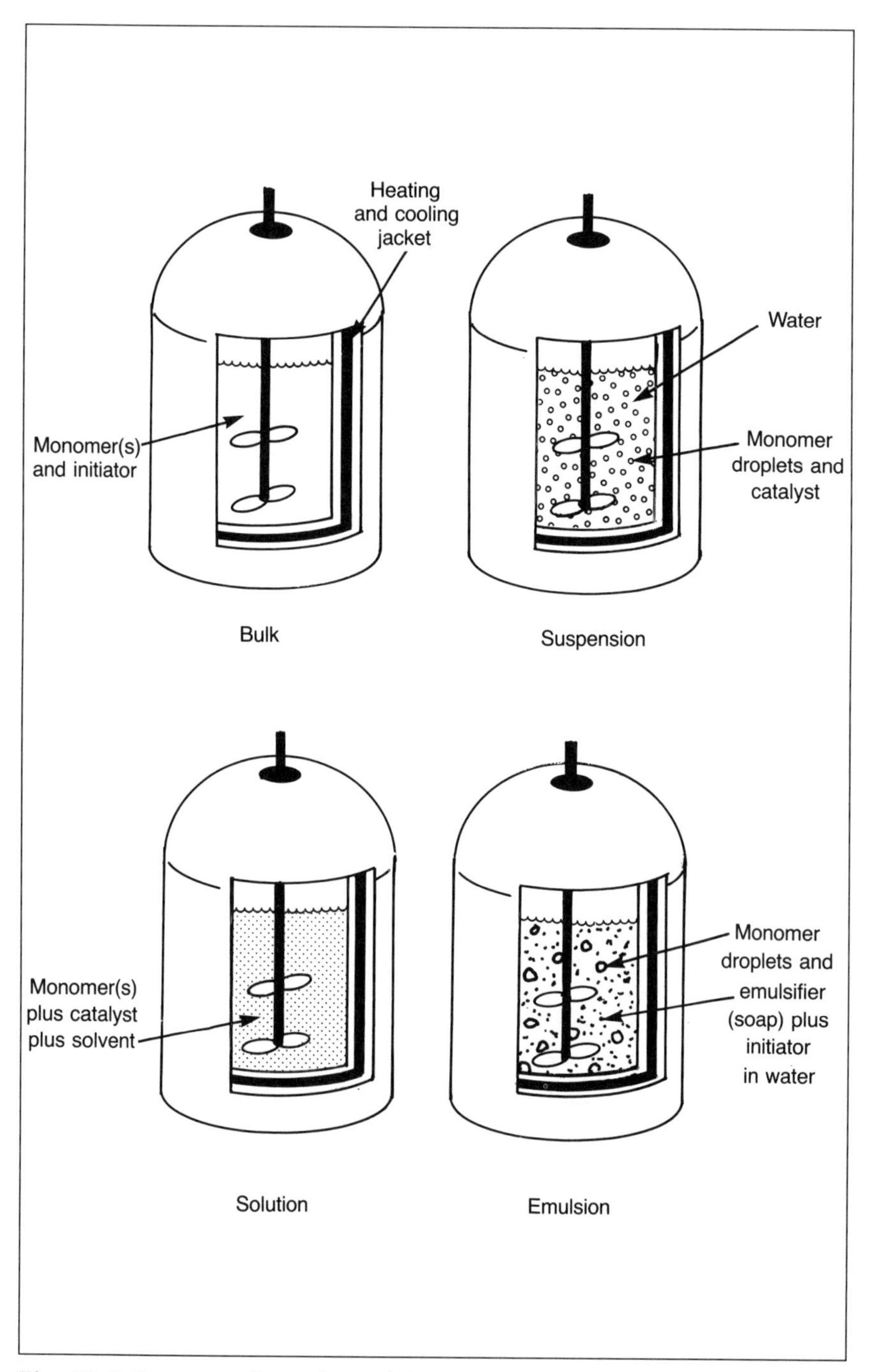

Fig. 20–5 Processes for polymerizing

Emulsion Polymerization. Soap is usually the emulsifying agent. The most useful characteristic of soap is the way the soap molecules behave in contact with oil and water. One end of a soap molecule is oleophilic (oil loving) and the other is hydrophilic (water loving). In an oil/water solution, the soap molecules form micelles, tiny structural units, suspended in the water. The oleophilic ends are directed inward, holding onto some (oily) monomer. The hydrophilic ends are pointed outward, interacting with the water medium. The polymerization actually takes place within the micelle, which remains suspended in the water. So, as it grows, the polymer remains suspended in the water. Very high molecular weight polymers are produced by this technique in the form of a latex. This process is particularly suitable for polymers used in paints, like polyvinyl acetate.

Polymer Properties

The proof of the polymer is in its properties. It is the physical properties that the engineers use in selecting polymers. They include density, tensile strength, impact strength, toughness, melt index, creep (ability to elongate), modulus of elasticity, electrical characteristics, thermal conductivity, appearance, flammability, and chemical resistance. Add price and fabricating costs and the engineer has most of the data he needs to select the correct polymer for his application.

Generally speaking, the physical properties of polymers depend on crystallinity, molecular weight, molecular weight distribution, linearity/cross-linking, and chemical composition/structure.

Crystallinity is one of the key factors influencing properties. You can think of crystallinity in terms of how well a polymer fits in an imaginary pipe, as in Figure 20–6. Linear, straight chains are highly crystalline and fit very well. Bulky groups, coiled chains, and branched chains are not able to "line up" to fit in the pipe. They are amorphous, the opposite of crystalline. In a spectrum from totally amorphous to almost totally crystalline, there is methyl methacrylate, polypropylene, low density polyethylene, linear low density polyethylene, high density polyethylene, and nylon.

With increasing crystallinity, polymers tend to be more dense. They're not too different from pasta. A pound of uncooked spaghetti fits in a smaller box than a pound of uncooked macaroni. The spaghetti is like

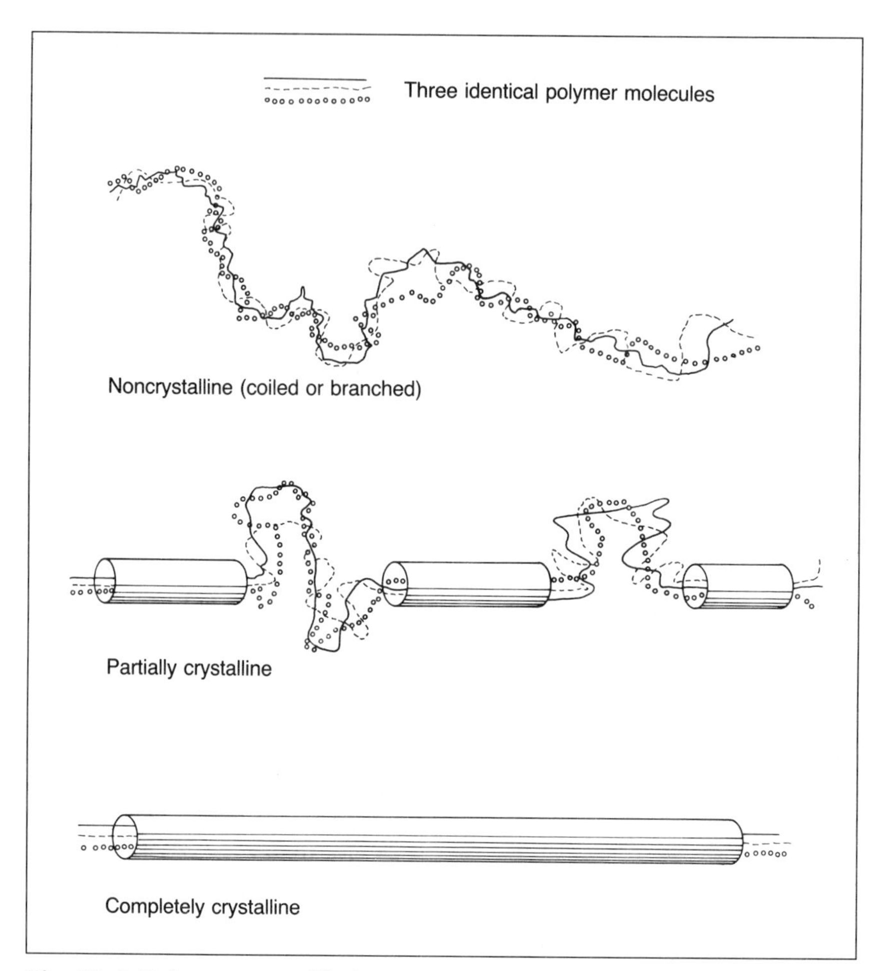

Fig. 20–6 Polymer crystallinity

a perfectly crystalline structure. The macaroni is the opposite. Along with increasing density comes greater tensile strength, higher softening point, and more opaqueness. Elongation (stretch) and impact strength decrease. The most surprising of these relationships—the more crystalline, the less translucent. It doesn't help to ask you if that's "crystal clear," because that expression implies just the opposite.

Molecular weight influences the melt viscosity, tensile strength, the low temperature brittleness, and the resistance to tearing.

Structure and chemical composition affect a number of the properties of polymers. These are discussed below.

Thermal Stability. The presence of side chains, cross-linking, and benzene rings in the polymer's "backbone" increase the melting temperatures. For example, a spectrum of polymers with increasing melting temperatures would be polyethylene, polypropylene, polystyrene, nylon and polyimide.

Stress-Strain Characteristics. Linear chain polymers are quite flexible and subject to creep or stretch. Branching or rings in the "backbone" have a stiffening effect. For example:

Polyethylene	—soft, tough, high creep
Polypropylene	—hard, tough, medium creep
Polystyrene	—hard, brittle, low creep
Cross-linked Thermosets	—hard, brittle, no creep

Density. Once something more than C-H is introduced to polymers, most of them get more dense. In order of increasing density are polypropylene, polyethylene, polystyrene, polyvinyl chloride, and Teflon.

Flammability. Presence of chlorine, fluorine, bromine, or phosphorous in a polymer reduces flammability. Thermosets are more flame resistant than thermoplastics.

Moisture Absorption. Directly related to the atoms making up the polymer. The more moisture-absorbing the molecule, the less dimensional stability; strength, stiffness, electrical properties are also adversely affected.

Those are the generalities of polymers. The specifics of low and high density polyethylene, polypropylene, polyvinyl chloride, and polystyrene are covered in the next chapter—resins and fibers in the last.

Chapter XX in a nutshell...

Polymers are high molecular weight compounds made by joining together hundreds or thousands of molecules. These molecules usually consist of one or two types of petrochemicals called monomers.

Polymers are generally one of two types.

—Thermoplastics: These can be dissolved or softened and remolded several times after their initial production.

—Thermosets: These set permanently and cannot be remolded, and are not meltable or soluble.

Thermoplastics are generally long chain, linear, two dimensional molecules, while thermosets are generally three dimensional long chains, connected by cross-linking chemical bonds.

There are two different types of chemical reactions used to make polymers:

—the addition reaction, where monomers are added end to end like, for example, polyethylene or polystyrene.

—the condensation reaction, where two or more different kinds of monomers are used to form copolymers. They are always formed by chemical reactions involving two reactive sites one on each monomer. Some by-product like water, hydrogen chloride is always formed. An example is ethylene glycol and terephthalic acid to make polyethylene terephthalate (polyester fiber).

The properties of polymers vary considerably, making the match between polymer and application a sort through such characteristics as density, tensile and impact strength, toughness, melt index, creep, elasticity, heat and chemical stability, electrical properties, flammability, and price.

Exercises

1. What's the difference between a thermoplastic and a thermoset?
2. What are the four basic steps of addition polymerization?
3. Name four process generally used for polymerization reactions.
4. Only __________ __________ can be used in making fibers
5. Which of the following polymer structures are homopolymers and which are copolymers? What monomers are involved?

   a. $-CH_2-O-CH_2-O-CH_2-O-CH_2-$

   b. $-O-CH_2-CH_2-O-\overset{\overset{O}{\|}}{C}-CH_2-CH_2-CH_2-CH_2-\overset{\overset{O}{\|}}{C}-$

   c. $-\underset{}{\overset{\overset{Cl}{|}}{CH}}-CH_2-\overset{\overset{Cl}{|}}{CH}-CH_2-\overset{\overset{Cl}{|}}{CH}-CH_2-$

THERMOPLASTICS

"Let there be
spaces in your
togetherness."

■

The Prophet
Kahil Gibran, 1883-1931

Like it or not, it's a plastic world out there. Plastics have penetrated the traditional markets for paper, cotton, wool, wood, leather, glass, metals, and concrete. (It's a good thing you can't eat it.) The growth of plastics would be even faster if they weren't made out of such an expensive raw material, petroleum. But many of the materials they are replacing have important energy components in their creation as well. So the advances in plastics continue.

In this chapter, the big four thermoplastics are covered, polyethylene, polypropylene, polyvinyl chloride, and polystyrene. Like most other thermoplastics, they are long chain polymers that become soft when heated and can be molded under pressure. They are linear or branch chained, with little or no cross-linking. Technological advances continue. Research in copolymerization, catalysts, processing, blending, and fabricating are continuing even as you read this.

POLYETHYLENE

You have to shake your head in wonder when you think about how the largest selling plastic was developed—by accident. In 1933, the scientists at the ICI labs in England were attempting to make styrene by the high pressure reaction of benzaldehyde with ethylene. Instead, they ended up with a reactor lined with a solid, white, wax-like material—polyethylene (Fig. 21–1). Six years later, a German scientist at IG Farben-Industrie, Max Fischer, was attempting to synthesize lube oils from ethylene. He tried a catalyst of aluminum powder and titanium at low pressures and ended up with a solid, white, wax-like material—polyethylene again.

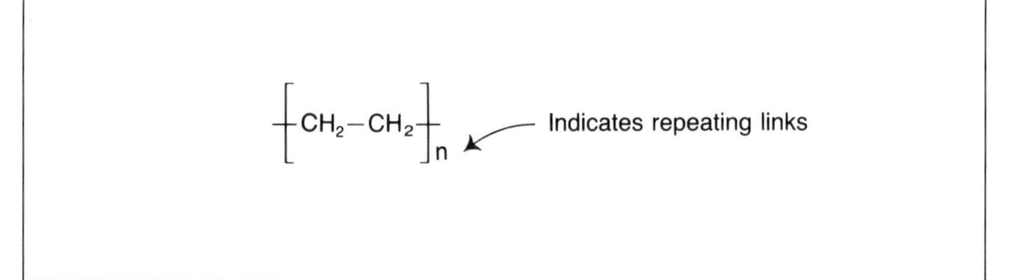

Fig. 21–1 Polyethylene

The English experience eventually developed into the high-pressure polymerization route to Low Density Polyethylene (LDPE). The German experiment was the forerunner of the low pressure route to High Density Polyethylene (HDPE).

The next polyolefin to arrive on the commercial scene was polypropylene in 1957, but only after some conscientious, "on-purpose" research. The most recent arrival was Linear Low Density Polyethylene (LLDPE) in 1977. LLDPE combined some of the best features of both LPDE and HDPE by using a comonomer, butene-1, hexene-1, or octene-1.

The most important thing about polymers is properties: how they look, how they react and how they perform when you do things to them or with them. That's why different types of polyethylene were commercialized. What are the primary differences between LDPE and HDPE? LDPE is more flexible and has better clarity; HDPE has greater strength and less creep and is less permeable to gases. That seems to go along with density differences, but also has to do with molecular weights, branching, and crystallinity as you'll see below.

LLDPE has most of the good features of both LDPE and HDPE—strength, flexibility, clarity, good dielectrics, and high/low temperature stability (for wire and cable shielding).

Despite the names, there's really not much difference in the densities of these polyolefins. For each of them, the density varies according to the degree of polymerization generated in the process. But in general, LDPE is about 0.920–0.935 grams per cubic centimeter; HDPE is about 0.955–0.970 g/cc; LLDPE varies between 0.920–0.950. That's a variation of less than 5%. So molecular weights and chemical structure also influence properties of the three different polyethylenes.

The routes to the polyethylenes fall into several categories:

1. High pressure polymerization, using the free radical catalysts discussed in the last chapter, to give LDPE. Reactor pressures run as high as 50,000 psi, temperatures up to 500–650 °F.
2. Medium pressure polymerizations with metal oxide catalysts, such as chromium oxide or molybdenum oxide to give HDPE.
3. Low to medium pressure polymerizations with Ziegler catalysts to give HDPE. Ziegler catalysts are trialkyl aluminum/titanium tetrachloride, the not-your-everyday-chemicals developed as catalysts by Karl Ziegler in the 1950s.
4. Low pressure, gas phase, fluidized bed (the catalyst is fluidized along with the ethylene) polymerization of ethylene with a little comonomer, producing LLDPE.

Ironically, the high pressure process produces a low density product; low and medium pressure produce a high density material. You'd think it would be just the opposite. But it has to do with branching and crystallinity. The high pressure leads to less crystalline molecules. The less crystalline, the less dense. (Recall the pasta example in Figure 20–6. Uncooked spaghetti is more dense than uncooked macaroni, and the spaghetti-shaped polymers are completely crystalline.)

LDPE Process

Since the polymerization process for LDPE requires pressures of 15,000 to 50,000 psi, the equipment is very expensive. Even worse, the reaction is exothermic and is liable to run away, which could result in an explosion. As a matter of fact, the original developers of LDPE, the ICI Laboratories, shut down their development work for several years after their first on-purpose attempt to make polyethylene exploded and destroyed their lab.

A flow diagram is shown in Figure 21–2. Compressing polymer grade ethylene (99.9% purity) to the reaction pressure is a major step. Several stages of compression are necessary. Since ethylene will start to polymerize on its own above 212 °F, cooling in between compression steps is necessary. (Compression always causes gas temperature to increase.) The compressed ethylene and catalyst enter a vessel called an autoclave reactor. An autoclave is any vessel that can be closed up and can maintain pressure. Usually something goes on inside that generates pressure, like a chemical reaction. A doctor or dentist sterilizes his instruments in his office by means of a steam-generated autoclave. An autoclave reactor is generally designed to handle both high temperatures and pressures.

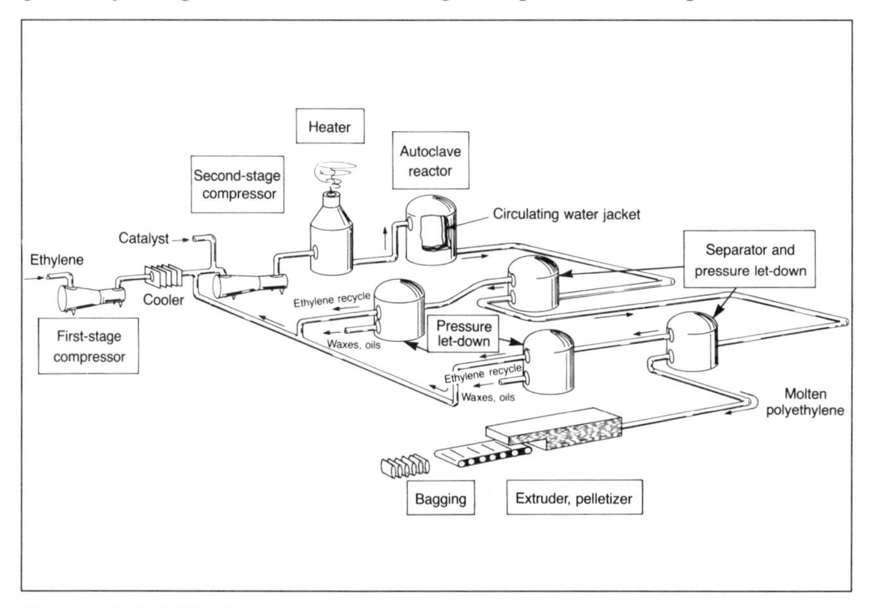

Fig. 21–2 LDPE plant

In the reactor is a mixture of ethylene and growing polymers. The polymerization reaction is exothermic enough so that the rate of ethylene added can be used to keep the reactor temperature constant. At the same time, polyethylene is constantly being drawn off to maintain the balance. This is called running the process under essentially adiabatic conditions (no heat needs to be added). But the water jacket around the autoclave reactor still is needed to act as a big sponge, sopping up excess heat that variations in the reaction can cause. It's also insurance in case of a runaway.

Residence time of the ethylene in this scheme is 25–30 seconds, on the average. Ethylene conversion, i.e., the amount of ethylene that gets used up, is only about 15 to 20% per pass.

To separate the ethylene in the effluent from the LDPE, the pressure is let down in successive vessels and the ethylene flashes off (vaporizes). The ethylene is recycled to the compressors. The LDPE, still in a molten (hot liquid) stage, is cooled, extruded, pelletized, dried, and bagged.

Often mineral oil is used as a carrier for the catalyst. Generally, both are left entrained in the LDPE. The mineral oil oil ends up acting as a plasticizer.

HDPE Process

In the 1950s, three different processes were developed independently to make HDPE (see Figure 21-3):

1. Solution
2. Slurry or particle form
3. Gas phase.

Solution process. In this route, the polymerization takes place in a vessel filled with a solvent like cyclohexane (Fig. 21-4). The solvent has several jobs. It keeps the reactants fluid, even after the polymerization; it sponges up a lot of the heat from the exothermic reaction; and it helps control the rate of ethylene consumption.

The reaction is also run in an autoclave reactor. Temperatures of 275–350°F and pressures of 300–450 psi are maintained. Residence times in the reactor of 1.5 hours are common — it's a slow process.

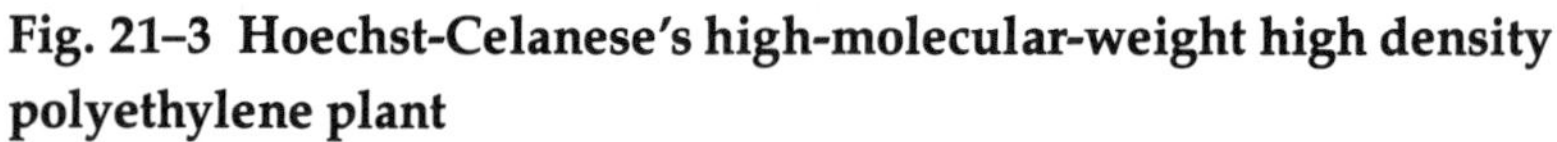

Fig. 21–3 Hoechst-Celanese's high-molecular-weight high density polyethylene plant

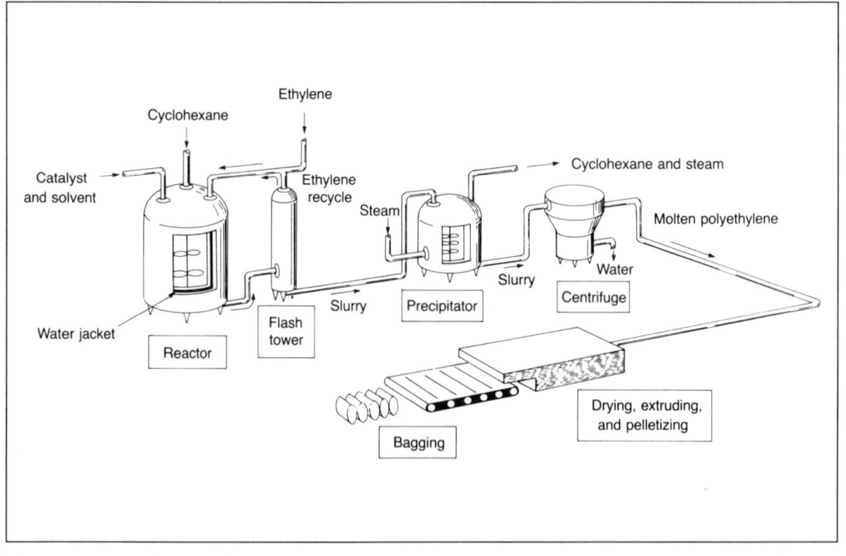

Fig. 21–4 HDPE solution process

Only small amounts of catalysts are needed, so catalyst separation from the reactor effluent often is not necessary. Downstream of the reactor are purification, recycle, and clean-up steps, until the polymer particles, called crumbs, are dried. The crumbs are then extruded, pelletized, and packaged.

The catalysts used in the reactor, chromic oxide, molybdenum oxide, or trialkyl aluminum/titanium tetrachloride (the Ziegler catalysts), are all very touchy about contaminants such as water, oxygen, peroxides, or acetylene. For that reason, the ethylene feed must be polymer grade (99.9% purity). The recycle streams, especially the cyclohexane solvent which has been through a steam wash, must be thoroughly dried. It doesn't take much more than trace amounts of those impurities to "poison" the catalysts.

Particle form process. The unique-looking equipment in Figure 21–5 is a loop reactor. This process also takes place in a solvent (in this case, it's usually normal hexane) so that the mixture can be pumped continuously in a loop while the polymerization is taking place. Feeds (normal hexane, comonomer if any, ethylene and catalyst) are pumped into the loop and circulated. Polymerization takes place continuously, and a slurry of HDPE in hexane settles in the vertical legs and is drawn off continuously or intermittently.

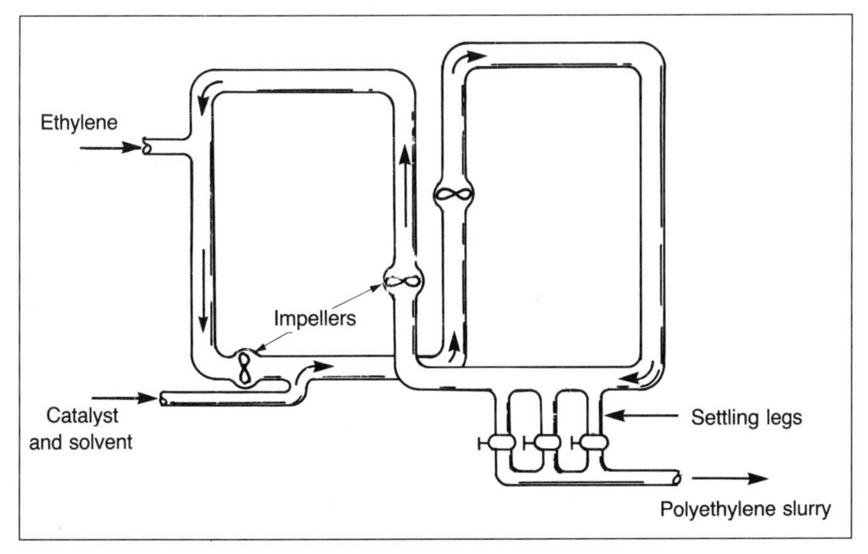

Fig. 21–5 HDPE particle-form process

The loops are pipes of 10 to 20 inch diameter, about 50 feet high, with a total length of 250–300 feet. They hold about 600 cubic feet of slurry and are water jacketed to control the heat. The reaction temperature in the process is less than 212°F, with pressures of only a couple of hundred pounds. So the process is more economical (energy saving) than the others already discussed.

Because of the longer residence times, this type reactor can make polyethylene molecules of higher molecular weights which also have high melt temperatures.

After the slurry is withdrawn from the loop reactor, processing is the same as that downstream of the reactor in Figure 21–2.

Gas Phase. No solvent is used in this process. Ethylene and a very reactive, silica-supported chromium-based catalyst are blown into a tall reactor. The ethylene and comonomer, if any, polymerize, drop to the bottom of the vessel, and are drawn off. The unreacted ethylene goes out the top of the vessel and is recycled. The HDPE is easy to clean up, since only the residual ethylene has to be separated. Coupling that with low temperatures (185–212°F) and pressure of 325 psi makes the gas phase method relatively inexpensive.

LLDPE

You might think of LLDPE as the third generation polyethylene. First there was the high pressure reaction for making LDPE in the 1930s. Then there were the low pressure HDPE processes in the 1950s, made possible by the Ziegler catalysts. In 1977, a process was introduced which permitted low pressures and temperatures, but the polyethylene was still low density (0.92–0.96 g/cc). The technological breakthrough was the use of the metal catalysts and a comonomer which permitted the use of low pressures but still turned out a low density polyethylene. About 8–12% of C_4, C_6, or C_8 alpha olefin is used as a comonomer. The hardware is much like the low pressure gas phase route for HDPE.

POLYPROPYLENE

When polypropylene (PP) technology finally ripened in the late 1950s, the chemical industry was quick to harvest numerous applications. The primary attractions of this thermoplastic were the ease of molding or extruding it and its ability to hold color. Some of the familiar applications are automotive parts, luggage, pipe, bottles, fiber (particularly carpet face fiber and rope), housewares, and toys.

Some special problems arise in explaining PP, and they breed a set of new terms that are used throughout discussions on any polymer more complicated than polyethylene. The problem lies in the extra group that propylene carries along. Except for that methyl group, $-CH_3$, propylene (CH_3-$CH=CH_2$) would be ethylene. Another problem is the "allylic" hydrogens* on the methyl group, which are reactive and capable of being displaced. (See Fig. 21–6.) That can lead to branching and sometimes cross-linking, which would affect the polymer's properties. As a matter of fact, the difficulty of controlling branching and cross-linking held up commercial development of PP until 1952. Then an Italian chemist, Giulio Natta, used Karl Ziegler's catalyst to produce a propylene polymer that finally had some useful properties.

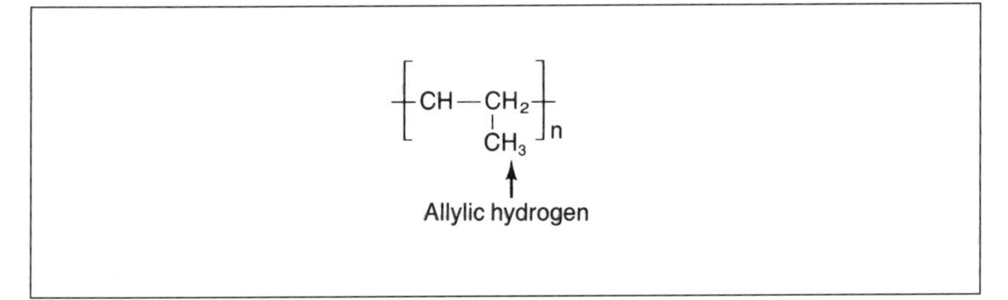

Fig. 21–6 Polypropylene

Understanding the chemistry of PP requires you to know the critical difference between PP and the polyethylenes—the asymmetry of the PP molecule's "backbone." In polyethylene, every carbon looks like every other carbon in the chain. In PP, the polymer linkage is between succeeding double-bonded carbons, like polyethylene. But, the methyl group survives as a branch on every second carbon in the PP "backbone" chain (See Figure 21–7.) Furthermore, the orientation of that branch is crucial to the properties of the polymer.

There's a whole area of chemistry dealing with the spatial configurations of organic molecules called stereochemistry. To get into this area,

*Allylic, from the Latin allium, meaning garlic. The $-C_3H_5$ is radical found in garlic and mustard.

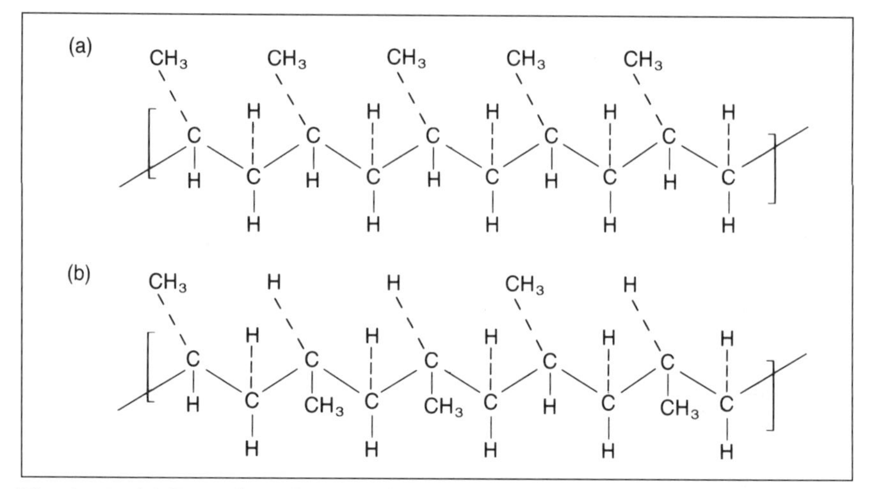

Fig. 21–7 (a) Isotatic polypropylene (methyl groups in the same plane); (b) Atactic polypropylene (methyl groups randomly in and out)

you have to have molecules that have an asymmetrical carbon atom. That's one that has four dissimilar atoms or groups attached to it. PP has that condition on a repeating basis—the methyl groups on every other "backbone" carbon. Such a polymer can be stereoregular or stereospecific.

In PP, stereoregularity of the methyl group is important. It really makes a difference whether every one sticks out in the same direction (more accurately, same plane). There are three possibilities that have been identified in PP molecules.

1. Isotactic: All the methyl groups are in the same plane.
2. Syndiotactic: The methyl groups are alternately in the same plane.
3. Atactic: The methyl groups are randomly in and out of the plane.

Of these three PP isomers (called that because they all have the same formula, just different stereoconfigurations), isotactic makes the best plastic. Atactic polypropylene is soft, elastic, and rubbery but not as good as rubber, natural or synthetic. It is usually separated from the isotactic propylene and discarded as waste, which adds considerable cost to the remaining isotactic. The isotactic form has a high degree of crystallinity with the chains packed closer together (makes sense—the molecules are more regularly oriented). The greater crystallinity gives higher tensile strength, heat resistance, dimensional stability, hardness, and a higher melting point.

So you can see why branching in the polymerization process can be a problem—the symmetry is affected. And you can get a hint why PP was commercialized long after polyethylene. The chemistry and catalysis are a lot more demanding. That's why Giulio Natta won the Nobel Price for his contribution to the field of stereo-catalysis.

The PP Plant

The facilities for making PP are very similar to those for HDPE in Figures 21-2 and 21-4. In fact, the same plants can be used. Like the stringent specs for polyethylene plant feeds, propylene must be polymer grade (99.5% or higher). Water, oxygen, carbon monoxide, or carbon dioxide will poison the Ziegler-Natta catalyst.

The only difference from the HDPE processing steps is the addition of facilities to remove atactic and low molecular weight PP from the isotactic, higher molecular weight polymer. This is done by making a slurry of the polymer mixture in hot normal heptane. The low molecular weight polymers and atactic PP will dissolve. The isotactic won't and can be removed from the slurry by centrifuging.

Process improvements are still underway, mainly in the catalyst area. The objective is to get the percent of isotactic polypropylene to approach one hundred, minimizing the atactic form. While most operations are now producing 60-70% isotactic, the newer catalysts give almost 95%.

POLYVINYL CHLORIDE

Leaf and grass bags, phonograph records, automobile upholstery, drainage pipe, roofing, and siding are all made from polyvinyl chloride. PVC is *the* consumer product plastic.

Vinyl chloride polymers and copolymers are often referred to as "vinyl resins." Originally they were based on acetylene. The switch to ethylene chemistry came after the development of the oxychlorination process for vinyl chloride described in Chapter IX. Today very little acetylene-based vinyl chloride monomer (VCM) processing remains.

PVC is the most important member of the vinyl resin family, which includes polyvinyl acetate (PVAC), polyvinyl alcohol (PVA), polyvinylidine chloride (PVdC) and polyvinyl acetal. Usually the term PVC includes polymers of VCM as well as copolymers that are mostly VCM.

Plasticizers

PVC (the homopolymer) is rarely used alone. Usually additives and plasticizers are added, more so than any of the other major thermoplastics. Plasticizers act like inter-molecule lubricants and can change pure PVC (and other polymers also) from a tough, horny, rigid material to a soft and rubber-like one.

Tricresyl Phosphate (TCP, the popular gasoline additive in the 1960s) used to be the popular plasticizer, but dioctyl phthalate has now replaced it. Dioctyl phthalate is the primary end-use for 2-ethylhexanol.

Plasticizer is usually added to the polymer during the compounding stage—that is, when it's being readied for molding, extruding, or rolling. The plasticizer is added in a hot mixer or roller operation. If PVC is expected to be plasticized, the polymerization steps can be controlled to produce a polymer particle that's very porous. Typically, for a flexible PVC, 25 to 30% of the finished PVC weight is plasticizer.

Adding plasticizer like dioctyl phthalate is generally accomplished by mechanical methods. Permanent or chemical plasticization can be done by copolymerization of VCM with monomers such as vinyl acetate, vinylidine chloride, methyl acrylate, or methyl methacrylate. Comonomer levels vary from 5 to 40%. The purpose of the co-polymers, of course, is to change the properties such as softening point, thermal stability, flexibility, tensile strength, and solubility.

Another way to vary PVC properties is to mechanically or physically add in other polymers such as ABS, SAN, MMA and nitrile rubber. These mixtures will improve the processibility and the impact resistance of the rigid PVC products.

Manufacturing PVC

Like polypropylene, PVC has the problem of stereospecificity. The carbon atom to which the chlorine atom is attached is asymmetrical. (See Fig. 21–8). As a result, PVC molecules can be isotactic, syndiotactic, and atactic. Commercial PVC is only 5 to 10% crystalline—low percent isotactic. It is more dense, 1.3 to 1.8 g/cc, than the polyolefins.

VCM can be polymerized by all four processes: suspension, emulsion, bulk, and solution. Most PVC is made by the suspension method because the polymer is more suitable for molding, extruding, or calendering (that's calender, with an -er, which means rolling into thin sheets).

PVC in a latex form comes from the emulsion process, the second largest route. PVC latex can be used for coatings "as is" or can be made ready for molding. In that case, it is spray dried, and the PVC particles are put into a liquid plasticizer (called plastisol) or into a mixture of plasticizer and organic solvent (organisol). The PVC particles do not dissolve but remain dispersed until the mixture is heated. Fusion then occurs, yielding the final plastic object. This is useful in forming special shapes by loading a mold with the plastisol or organisol and heating. (Vaporization of the organic solvent is often used to create a foam. See Foams below.)

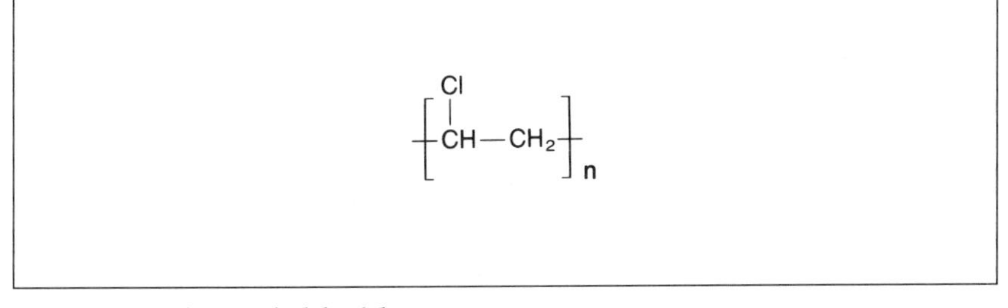

Fig. 21–8 Polyvinyl chloride

The PVC Plant

In the suspension polymerization process, the autoclave reactor is filled with water. Polyvinyl alcohol is the agent that helps stabilize the suspension. Lauroyl peroxide is the free radical catalyst that starts it all off. The reaction temperature is around 130 °F, and the process takes 10–12 hours per batch, with 95% conversion.

The reactors are typically 5000–6000 gallon, glass-lined, water jacketed vessels. (See Figure 21–9.) After all the ingredients are loaded in, steam is run through the jacket to get the mixture up to 120–150 °F. After the reaction begins, cooling water replaces the steam in the jacket to take away the heat generated in the exothermic process. Meanwhile, the vessel contents are mixed vigorously to keep the monomer suspended in the water. The PVA helps out here. The polymer molecules have to keep bumping into each other to keep growing.

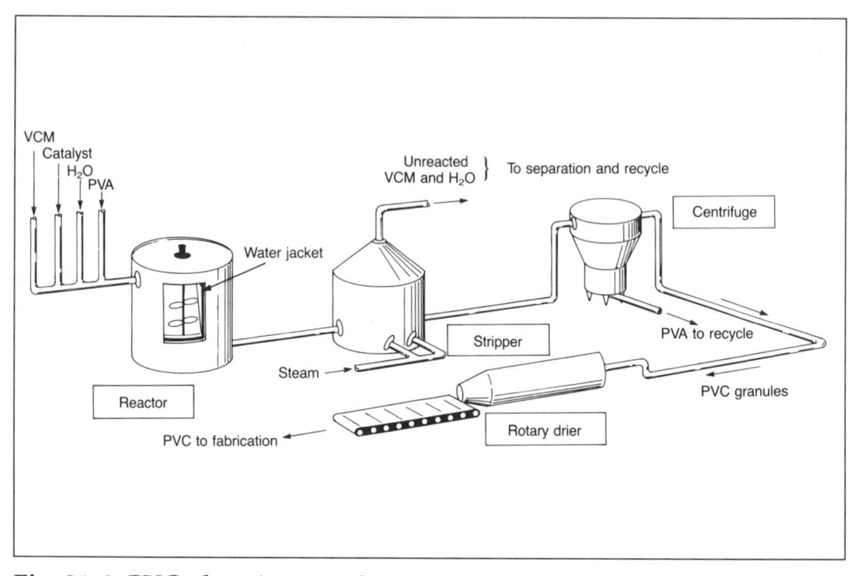

Fig. 21–9 PVC plant (suspension process)

To get a porous PVC bead that will accept high levels of plasticizer, a sudden pressure release is sometimes used. During the 10–12 hours of cooking, the reactor pressure will increase as the temperature goes up, then decline slowly as the polymerization approaches 100%. If the pressure is suddenly released during the process, some of the unreacted VCM will vaporize. The granules of PVC that have formed will begin to swell as it absorbs the VCM vapor. As the vessel is "buttoned back up" and repressured, the VCM will reliquefy, but the PVC already formed remains swollen and porous.

After the reaction is completed, the suspension is transferred to a degassing tank, where steam is used to strip out the unreacted VCM. The PVC, now in a slurry with PVA, is separated by centrifuging and dried. The PVC powder or granules are then ready for additives and plasticizers for fabrication into one of the three mediums in which it's used: calendered products, extrusion, and molded products.

POLYSTYRENE

When you hear polystyrene (PS) you probably think of products made of PS foam—disposable coffee cups, packing materials, buoys and boat bumpers, and cheap ice chests. As a matter of fact, PS foam products are so important that there's a special section at the end of this chapter dealing with foam.

Foam accounts for less than half of the polystyrene output. The remaining products have properties very different from foam. PS is an excellent plastic for molded automobile and refrigerator parts. It accepts color so well that it is widely used in molding applications to simulate wood. Probably all the "wood" on your new console TV is PS.

There's a lot of competition between PS and the other five big thermoplastics: LDPE, LLDPE, HDPE, PP, and PVC. Polystyrene continues to lose market share, but it seems to have a permanent place in some applications, particularly molded foams (for carry-out food containers), some extrusions, sheet and film applications.

Manufacturing PS

Like PP and PVC, each repeating monomer unit (Fig. 21–10) in PS has an asymmetric carbon atom. It's the phenyl group (benzene ring) attached to this carbon atom that makes the polymer asymmetrical. The polymer can be iso-, syndio-, or atactic. Commercially produced PS is usually an atactic amorphous polymer (low crystallinity, good optics). The isotactic form can be made using the Ziegler-type catalysts. However, there's no major, marketable improvement in its properties so most processes produce the cheaper atactic form.

All four polymerization processes can be used to make PS. The reaction is an addition polymerization using a free radical initiator (benzoyl peroxide or di-tertiary butyl peroxide). Mostly, the suspension or bulk processes are used. The suspension process is identical to the PVC process shown in Figure 21–9. As just one more mind expander, the bulk process will be covered here.

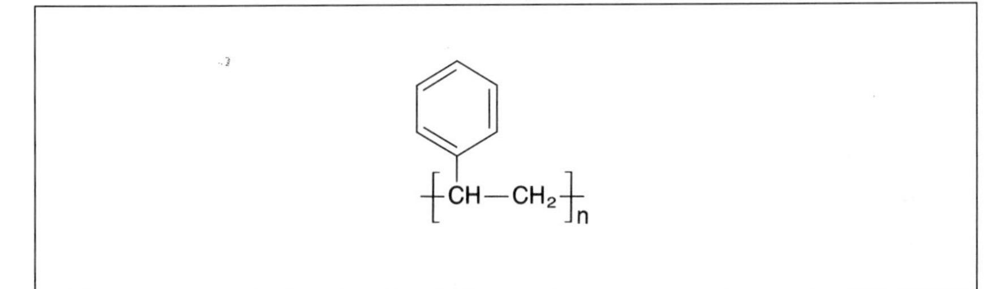

Fig. 21–10 Polystyrene

The Polystyrene Plant

The bulk polymerization process needs monomers that can dissolve their own polymers. (There's no solvent or water in the reactor to keep the polymer floating around.) Polystyrene and some of its copolymers have this property, and so it's generally cheaper to use bulk polymerization.

The process begins in a prepolymerizer, which is a water-jacketed reactor with a mixer in it (Fig. 21–11). The styrene is partially polymerized by adding the peroxide initiator and heating to 240–250 °F for about four hours. About 30% of the styrene polymerizes, and the reactor contents become a syrupy goo. That's about as far as the prepolymer step can go—30% conversion—because the mixing and heat transfer gets very inefficient as the goo gets thicker, and the polymerization becomes hard to control.

The goo is then pumped to the top of a vertical, jacketed tower with internal temperature-regulating coils. The vessel is kept full of the styrene/PS mixture. A temperature gradient (change) of 280 °F at the top to 400 °F at the bottom is maintained. The temperatures are controlled to prevent runaway, but to permit 95% conversion of styrene to PS. As the polystyrene molecules grow, they sink to the bottom of the vessel and can be drawn off. The residence time in this vessel is three to four hours. The molten PS is extruded to strands, chopped into pellets, and bagged.

The most critical factor in this process is temperature control in the second reactor. The viscosity of the mixture top to bottom changes with temperature, but also with PS concentration. If hot spots develop because of the exothermic reaction, a runaway can occur. In that event, the batch must be immediately quenched, ruining it. Several process improvements (not shown in Fig. 21–11) include using agitators, solvents, and solvent removers.

Applications

About 25% of polymerized styrene is in the copolymer form. The largest volume copolymer is SBR (25% styrene, 75% butadiene-rubber), used for making tires, hoses, belts, footwear, foam rubber, rubber-coated fabrics, and adhesives.

ABS (30% acrylonitrile, 20% butadiene, 50% styrene) are tough plastics with outstanding mechanical properties. ABS is one of the few

Fig. 21–11 Sterling Chemical Company's styrene monomer plant.

plastics that combines both toughness and hardness. So the applications include ballpoint pen shells, fishing boxes, extruded pipes, and space vehicle mechanical parts. There's about 20 pounds of ABS molded parts in an automobile.

SAN (70% styrene, 30% acrylonitrile) has better heat and chemical resistance and is stiffer than PS. The optical clarity is not as good. SAN is used in a variety of houseware applications, particularly those things that will come in contact with food (chemical attack) and those that will end up in a dishwasher (heat attack). Coffee pots and throw-away tableware are good examples.

FOAMS

Foamed polymers are low density, cellular materials that contain bubbles of gas and are made in a variety of ways out of thermoplastics and thermosets. Their properties vary from rigid to flexible. The rigid foams are best known for their insulation properties (like in ice chests). The flexible foams are used extensively in cushioning (seats, mattresses).

The difference between rigid and flexible foams, from a simplified view, is the nature of the cells that make up the foam. Rigid foams are made up of closed cells. The gas they contain is sealed in. Flexible foams have open cells. When you compress a flexible foam, all the air can be squeezed out. When you compress a rigid foam, the gas cannot escape, so nothing moves. Simple, isn't it?

The closed cells also give rigid foams their excellent insulating properties. Gas is a notoriously poor conductor of heat. That's why storm windows work. They have a dead air space in the middle. Rigid foams are just like storm windows. They trap a dead air space. Flexible foams wouldn't do quite as well because they let the air move around.

Foams are commercially produced several ways. Some polymerization processes produce their own foam. Polyurethanes, for example, are very exothermic. When they are formed, if a little water is present, CO_2 will be a by-product. As the polymer forms, the CO_2 will cause closed cell foam. As another example, a blowing agent can be injected into the molten polymer. The agent will later decompose, giving off a gas when the polymer is heated to melting. Epoxy resins are expanded into foams this way.

A popular, related technique is to inject some sort of a volatile material into the polymer while it's still molten, causing it to foam immediately. Fluorocarbons are used this way in making polyurethane foams suitable for insulation. The trapped fluorocarbon is even better than air as an insulator. Fluorocarbon or air is used to expand PS foams also. The gas is injected as the molten polymer is forced through a die. The foamed PS is then immediately injected into a mold to make items like egg cartons and trays for meats, produce, or fast-food trays.

Expandable PS beads are a material devised to accommodate the transportation drawbacks of foams. Foams take up a lot of room, but not much weight, so a truck or box car cannot be used very efficiently. Expandable PS beads can be readily turned into foam at their destination. The beads are impregnated with a volatile liquid like pentane as they are extruded, chopped, and cooled. Later, on site, the beads are heated in small batches with steam. The vaporization temperature of the pentane is just below the melting point of the PS beads. As the beads soften, the pentane flashes (volatilizes) and causes the PS to foam. The polymer is then ready for molding. Coffee cups, ice chests, life preservers, buoys, and floats are often fabricated this way.

Most thermoplastics and thermosets can be foamed, many of them into either flexible or rigid foams. The choice is controlled by the blowing agent, additives, surfactants, and mechanical handling. Some polymers can be expanded as much as 40 times their original density, and still retain a substantial part of their strength. Most commercial foams are expanded to densities of two to five pounds per cubic foot. (Water is 62 pounds per cubic foot.)

PLASTIC PROPERTIES

That worn-out joke applies: "The three most important things about plastics are (1) properties, (2) properties, and (3) properties." It's impractical to cover all the dimensions, but here are some of the most important.

PE has excellent electrical properties, good clarity, good impact strength, and is translucent in thick sections. It also has good chemical resistance and excellent processibility.

PP is the lowest density plastic. It has fair-to-good impact strength and excellent colorability. It's translucent in thick sections, and it also has good chemical resistance. The properties can vary widely with different degrees of crystallinity. PP has good resistance to heat and low water absorption. That makes it a suitable material for many medical instruments which need sterilization by steaming.

PVC has good electrical properties and is flame resistant with the proper plasticizers. It's even self-extinguishing. It has good impact strength and chemical resistance. Although rigid "as made," it is easy to make flexible by the addition of plasticizers. It does require heat and light stabilizer additives.

PS is easily processible and has excellent color, transparency, rigidity, dimensional stability, good tensile strength and electrical properties. It is easily foamed.

Chapter XXI in a nutshell...

The six most popular thermoplastics, low density polyethylene, high density polyethylene, linear low density polyethylene, polypropylene, polystyrene, and polyvinyl chloride, are all linked in the same chemical configuration, via the ethylene group. The only difference is that three of them have a substituted group for one of the hydrogens belonging to ethylene, $CH_2=CH_2$. Polypropylene has a methyl group hanging off, polystyrene has a benzene group, and polyvinyl chloride has chlorine.

The combination of these unique appendages—plus the differ equipment, reaction pressures and temperatures, use of comonomers and catalysts—result in quite different properties for each thermoplastic. But they all can be remolded, melted, or dissolved a number of times after initial formation.

Exercises

1. What does the term copolymerization mean and give some examples?
2. What's the "bad" kind of polypropylene and why is it "bad?"
3. What are all the building blocks and intermediate petrochemicals that lead up to polyvinyl chloride?

RESINS AND FIBERS

"Tell her to
make me a
cambric shirt."

■

"Scarborough Fair" (1966)
Paul Simon and Art Garfunkel

After you've plowed through the thermoplastics, you only need to read about the resins and fibers to cover the rest of the applications for most petrochemicals. That's one reason for putting resins and fibers in one chapter. The other is that some polymers, like nylon, can be both a resin and a fiber. You just grow them a little differently.

The coverage in this chapter is compact—no detailed process descriptions or diagrams. Resins and fibers aren't really petrochemicals anyway. They're just a good climax to the petrochemical story.

RESINS

Chapter XX didn't give you a very satisfying definition of resins. But it's useful here to talk about two classes of polymers called resins: thermosets and engineering thermoplastics.

Thermosets

You'll recall that thermosets are polymers that have lots of cross-linking. The molecules are three-dimensional, rather than two. More importantly, once the cross-linking bonds are in place, the polymer becomes rigid and hard. Put another way, once the thermoset occurs, it is irreversibly set. That's the difference between thermosets and thermoplastics. The latter can be remolded and reshaped; the former cannot. When you sweep up the scrap material around the molding/extruding machines that handle thermosets, you throw it away.

Phenolic Resins. The oldest condensation reaction on record is between phenol and formaldehyde to produce phenolics. The reaction was first documented by Baeyer in 1872. Thirty years later a technical application of this reaction was worked out by Dr. Leo Bakeland, when he showed that useful moldings can be made by carrying out the final stages of the reaction under pressure. As his reward, phenolic resins are still often called "Bakelite." At one time, phenolics were the workhorse of the plastics industry.

The chemistry of phenolic resins is complex. Even today it is not fully understood. The brief description that follows is definitely an oversimplification of the reactions involved.

To make phenolic resins, you have to first make a phenolic prepolymer, which may be in a liquid or solid form. This can be accomplished by using either a base or an acid catalyst. The prepolymer is a low molecular weight, linear polymer that—and this is the whole key to phenolics—can be further processed at a time of the processor's choosing to give the cross-linked phenolic resin. All you need is a little more heat and pressure. The reactions are shown in abbreviated form in Figure 22–1.

Phenolic resins are the cheapest of all molding materials, since they usually contain more than 50% filler—sawdust, glass fibers, oils, etc. Their main properties are heat resistance, excellent dielectrics, and ease of molding. However, they have poor impact resistance (they crack), and

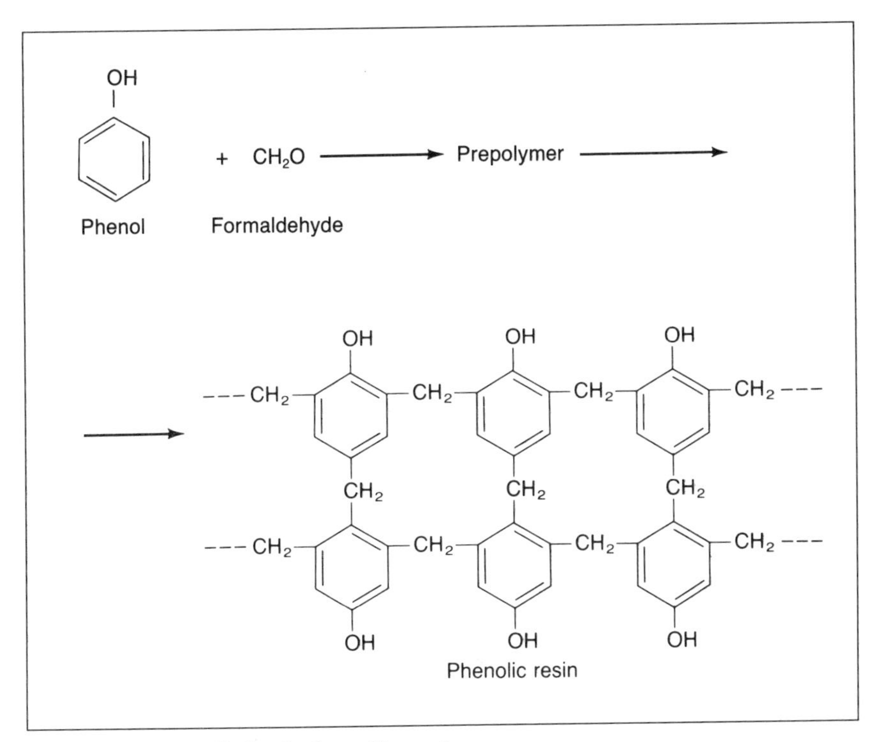

Fig. 22–1 Cross-linked phenolic resin

they don't hold most dyes very well, except black. Their use is thereby restricted—they're functional but not pretty. When the telephone companies started making phones in colors, they quit using phenolic resins and instead bought more expensive thermosets.

Other applications for phenolics are switchgears, handles, and appliance parts, such as washing machine agitators (that's why they're usually black). Phenolics are widely used to bond plywood, particularly exterior and marine grades. Although urea-formaldehyde resins are cheaper for this purpose, they are not nearly as water resistant and have been limited to interior grades. Abrasive wheels and brake linings also are bonded with phenolic adhesives.

Epoxy Resins. The reaction of Bisphenol A and epichlorohydrin gives a low molecular weight linear polymer. This polymer further reacts with an amine curing agent, R-NH_2, to give a general purpose thermoset (Fig. 22–2). Now that's not as complicated as it sounds. First of all, you'd think the name epoxy is used because the epoxy ring,

```
   O
  / \
- C - C-,
  |   |
```
first encountered in ethylene oxide, remains in the molecule. Well, it is and it isn't. It is in the prepolymer chain, which results from the Bisphenol A/epichlorohydrin condensation reaction. However, it is used up in the subsequent cross-linking step. This ring is the primary site where the polymer cross-links.

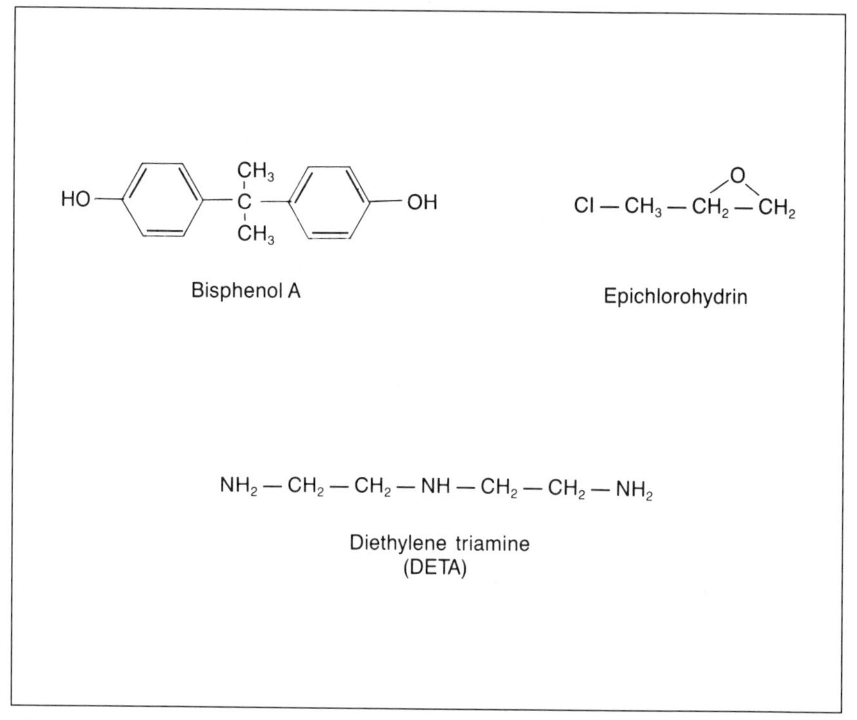

Fig. 22–2 Epoxy monomers

Epoxy resins are a post-World War II development. Initially, they were used as surface coatings. They are extremely resistant to heat and corrosion, and they have excellent adhesion to metals. Unfortunately, they get chalky when exposed to too much sunlight. Therefore, their use is limited to primers and places protected from sunlight, like coating pipeline interiors. Coatings are the largest volume use of epoxy resins, but the best known is the household adhesive that comes in two tubes. One contains the resin; the other has the "hardener," an amine catalyst plus filler. Common amine curing agents are diethylene triamine (DETA) and triethylene tetramine (TETA). The package is usually labeled "epoxy glue."

Other major uses are commercial adhesives, laminates, and potting for electrical components. Epoxy resins are particularly suitable for potting (imbedding electrical components in a non-conductive thermoset) because they provide dimensional stability (no shrinking) as the thermoset cures.

Polyurethanes. The word urethane has the same root word as urine, because both are related to urea, NH_2CONH_2. All three chemicals have the characteristic -N-C- linkage.

```
-N-C-
 | ||
 H O
```

Urethanes result from reactions between alcohols and isocyanates. So what's an isocynate? It's an organic compound having the -N=C=O signature grouping. Isocyanates are the product of an amine and phosgene. An amine is a compound with the formula $R\text{-}NH_2$. Phosgene is $COCl_2$, a product resulting from the reaction of carbon monoxide and chlorine.

Simple polyurethanes are made from diisocyanates and diols. The general form for a diisocyanate is O=C=N-R-N=C=O. The popular one, toluene diisocyanate (TDI), is shown in Figure 22–3. Diols are molecules that have two hydroxyl groups (alcohol signatures) attached, such as propylene glycol. Polymerizing TDI and propylene glycol gives a linear polymer because both molecules are bifunctional. That is, they each have only two sites where they can react, so they string out end to end.

If cross-linking is desired, as in a polyurethane foam, TDI can be reacted with a polyol or polyolether. These chemicals have multiple -OH groups that make them polyfunctional. Cross-linking can take place at one or more of the hydroxyl sites. Compounds other than TDI can be used to make polyurethanes. However, TDI accounts for about 60% of this production.

Polyurethanes are primarily found in flexible foams, but they also make rigid foams and laminates. Two methods are used to develop these foams. In one procedure, all of the ingredients are mixed, including a catalyst and a blowing agent additive, and put into a mold. The reaction kicks off immediately. Foaming is completed in a few minutes because the gases are liberated from the blowing agent. However, complete curing (cross-linking) takes several more hours.

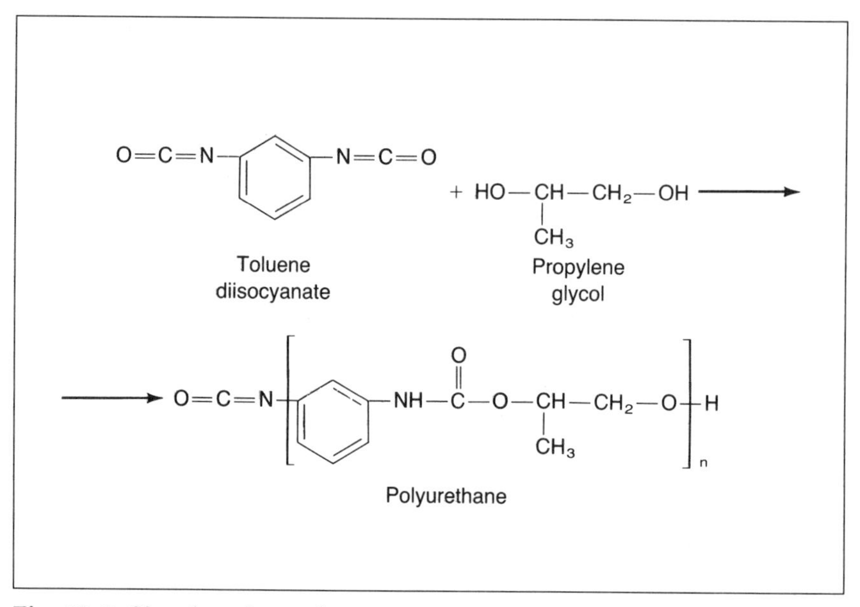

Fig. 22–3 Simple polyurethane

The other foaming process is referred to as the prepolymer method. The monomers are reacted to form a low molecular-weight prepolymer. Later the prepolymer is mixed with small amounts of water and is heated. The water reacts with the free isocyanate groups to liberate carbon dioxide, which foams the polyurethane as cross-linking starts.

Polyurethane foams are lighter than foam rubber and have displaced many of its applications, such as in bedding, cushions, car seats, arm rests, and crash pads. Laminates are used widely in clothing as padding.

Rigid foams are excellent insulators, even better than polystyrene, and are used in refrigerators and refrigerated trucks and box cars. Polyurethane coating materials are popular additives to marine finishes and varnishes, particularly for gymnasium floors, bar tops, and other surfaces that take an abusive, abrasive beating.

Amino Resins. Urea, the first recorded, synthetically produced, organic compound, can be reacted with formaldehyde to form polymers called urea-formaldehyde resins. Its chemistry is similar to the phenolic resins.

Articles made from amino resins are water clear, hard, and strong, but they can crack. The have good electrical properties, and they have better colorability than phenolic resins. Amino resins are used as adhesives for plywood and particle board but only in interior grades. They have low weather resistance and deteriorate when exposed to sun, heat, cold, and moisture.

Extensive use of these resins is found in textile and paper treating and surface coatings. Many types of clothing also can be given a permanent press by an amino resin treatment. Amino resins can be molded and are used for radio cabinets, buttons, and switchplate covers, dishware and formica.

Engineering Resins

Polymers with special properties (such as high thermal stability, good chemical and weather resistance, transparency, self-lubrication, and good electrical properties) can be called engineering resins. Nylon and polycarbonates are two good examples.

Nylon. The name nylon covers a number of polymer compounds, all of which are based on the amide linkage

$$\begin{array}{cc} -\mathrm{C} & \mathrm{N}- \\ \| & | \\ \mathrm{O} & \mathrm{H} \end{array}$$

shown in Figure 22–4. (Resins made from nylon actually account for much less volume than fibers from nylon.) The two most popular nylons, both in resins and fibers, are Nylon 6 and 66. These two account for about 80% of the production.

Nylon 6 is made by the addition polymerization of caprolactam. Caprolactam is a seven-sided hetrocyclic that should be drawn as a septagon, just like cyclohexane and benzene are hexagons. However, few people can draw septagons well, so most sketch caprolactam as a rectangle, like the one shown in Figure 22–5. The route to caprolactam is complex. It starts with cyclohexane. Three reactions later, out comes caprolactam.

Nylon 66 is made by the condensation polymerization of adipic acid and hexamethylenediamine (HMD). Adipic acid was covered in Chapter XVI (cyclohexane is also the building block for it), and HMD shouldn't be such a threatening word to you by now. The hex is six; the methylene is $-CH_2-$. The di is two, and amine is the signature group $-NH_2$.

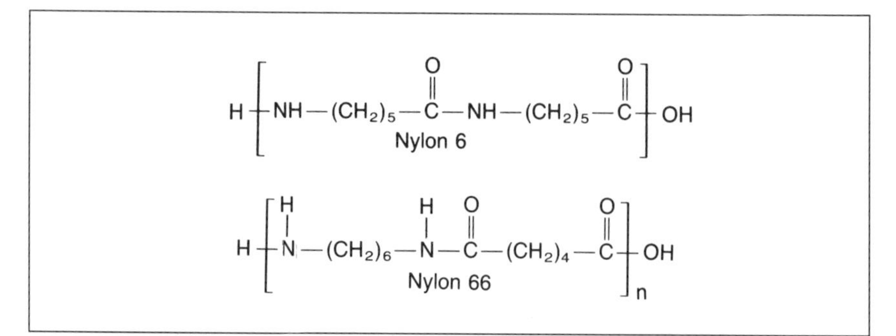

Fig. 22–4 Nylon

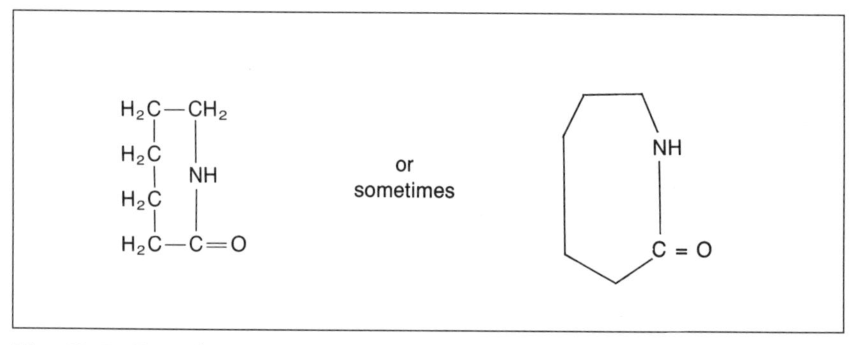

Fig. 22–5 Caprolactam

Put them all together and these groups spell $H_2N(CH_2)_6NH_2$, which is HMD. The routes to HMD and adipic acid are shown in Figure 22–6.

The 6 and 66 are part of an awkward numbering system used to indicate what nylon was made from. A number up to 12 indicates the nylon was made from a single monomer with that number of carbons in it. A number over 12 signifies that two different monomers were used, with the number of carbons in each adding up to the number designated. Check Figure 22–4 to see how it works. Examples of other nylons are Nylon 11, 12, and 610. Originally all double numbered nylons had a comma in the name—Nylon 6,6. Through constant misuse, the comma has mostly been dropped.

Polycarbonates. The polycarbonates surfaced in the 1950s, so they are relatively new polymers. They are made in a condensation polymerization process. The reactants are either Bisphenol A and phosgene or Bisphenol A, phosgene, and phenol. Since Bisphenol A is a derivative of phenol, the building block is the same in either case—phenol. The poly-

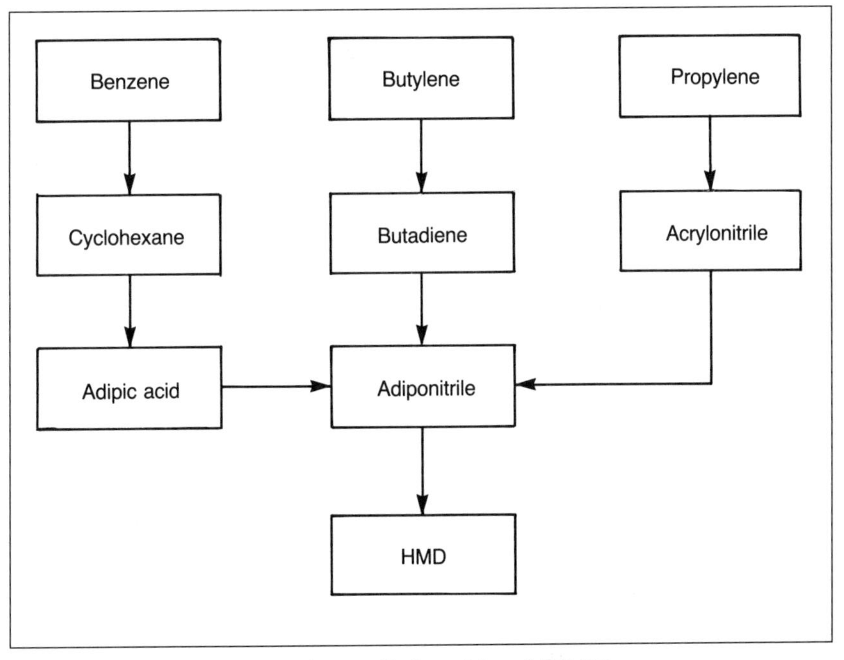

Fig. 22–6 Methods to produce adipic acid and HMD

carbonate based on Bisphenol A has the best balance of properties. If you look hard, you can see the monomers in Figure 22–7.

Polycarbonates differ mechanically from the epoxy resins they resemble chemically because polycarbonates are plastics. They can be molded or extruded from chips or crumbs. The products have high impact strength and can be used to make sturdy films and transparent forms. Applications are photographic film; goggles; helmets; streetlight fixtures; tool, machine, and appliance casings; and electrical and electronic compounds.

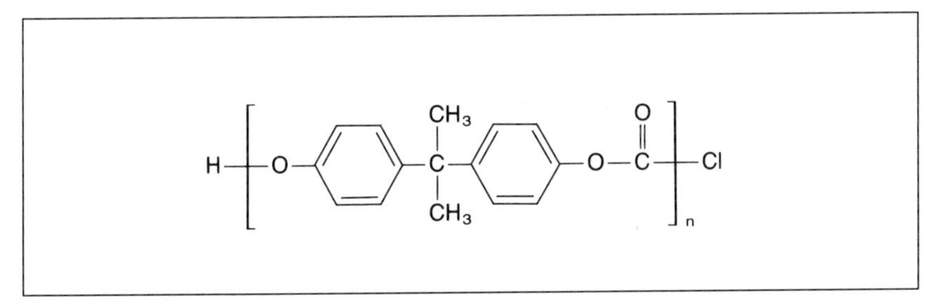

Fig. 22–7 Polycarbonate

FIBERS

Natural fibers go back to prehistoric days. Probably one of the early applications was the conversion of a fiber (possibly wool or cellulose) into thread or rope strong enough to be used in a snare, net, or cage. Literature as far back as the seventeenth century notes that people attempted to make fibers out of something other than cotton, wool, or flax. The first man-made fiber, known as artificial silk, was made in the nineteenth century, when wood pulp was treated with nitric acid. The result was known chemically as cellulose nitrate and (eventually) commercially as Rayon. The commercial name referred to the sheen which "has the brilliance of the sun."

However, Rayon isn't a petrochemical derivative. Those products did not emerge until the technological breakthrough of W. H. Carothers of DuPont in the 1930s. His classical research showed that chemicals of low molecular weight could be reacted to form polymers of high molecular weight. Some people consider Carothers the father of polymer chemistry. His original offspring was Nylon 66, the first commercial fiber to be made entirely from a synthetic polymer. (Some reports have PVC fiber beating Nylon 66 by two years, but PVC never really caught on as a fiber.)

The first Nylon 66 fiber production facility came on stream in 1939. The first nylon stockings were marketed in May 1940—just in time for G.I.'s to take them to Europe in World War II.

DuPont continued their leadership role in synthetic fibers by commercializing acrylic fibers (Orlon) in 1950. They did a repeat performance in 1953 with polyester (Dacron). The big four fibers—Nylon 6, Nylon 66, acrylics, and polyester—now account for most of the synthetic production and about half of the fiber production of all kinds, including cotton, silk, and wool.

The Mechanics of Fibers

The chemistry of fibers is the same as that for resins. The important difference is the mechanics. For polymers to be suitable for fibers, you must be able to draw them into a fibrous form, normally by extrusion. Second, the size and shape of the molecules that make up the fiber must be correct. To have acceptable fiber properties, the molecules must be long, so they can be oriented to lie parallel to the axis of the fiber. Normally that's done (or enhanced) by drawing or stretching the fiber to several times its original length. The essential differences then, between resins and fibers, are the shape and the orientation of the molecules.

Spinning is the classic process of twisting a bundle of parallel short pieces of natural fiber into thread. (You can't get more classic than a spinning wheel.) It also is the modern process for extruding long lengths of man-made fibers. In the classic process, the short lengths are known as staple fibers and the resulting thread is a spun yarn. The long, extruded lengths are called continuous filament or just filament yarn.

The denier of a fiber or a yarn defines its linear density. This term describes the weight in grams of a 9000 meter length of fiber at 70°F and 65% relative humidity. Denier generally denotes the size or diameter of the filament or yarn. Fibers usually range from 1–15 denier and yarns 15–1650 denier. Breaking tenacity is a measure of strength of a fiber or yarn and is measured in grams per denier. The stronger the fiber is, the higher the tenacity.

Nylon (Polyamide Fibers)

The chemical structure of the nylon fiber looks just like the nylon resin. The polymerization processes are the same; the numbering systems are the same; and the two most important nylon fibers are the same Nylon 6 and Nylon 66. The difference is the length of the molecule in comparison to the cross section. That's regulated by the polymerization process conditions.

After the polymer is formed, the filament is produced by extrusion. The nylon can be taken directly from the polymerization process while it is still molten. Alternatively, previously dried nylon chips can be heated and melted. The molten nylon is pressed through a spinneret, which is like a shower head with 10–100 holes. (A spinneret with a single hole produces a monofilament. Nylon monofilaments are used for fishing line and sheer hosiery.) The fine strands that form are immediately air-cooled, stretched, and wound on bobbins.

Then the fibers undergo cold drawing. Nylon 66 can be stretched 400–600% of its original length with very little effort. A very important process goes on during the stretch. The polymer molecules orient themselves to a controlled amount of crystallization, depending on the stretch conditions. This crystallization gives the fibers their special properties.

Finally the fibers are textured for specific applications. They can be twisted, coiled, or even randomly kinked, if they are to be used for carpet piling. More than half of the total carpet-fiber market is based on nylon staple and filaments. When nylon is to be blended with other fibers, the filament is cut into staple fiber—short pieces 3–15 centimeters in length—for subsequent spinning.

One of the larger uses of nylon fibers is tire cord. In apparel applications, which are another major area, permanent press can be achieved by heat treatment. This crease resistance lasts until abrasion, heat, or pressure wears down the molecule orientation. Since it is strong and lightweight, nylon also is used for rope, parachutes, and some undergarments.

Polyesters

The process for producing polyesters wasn't covered in the sections on thermoplastics or on resins. So read it carefully and it will make sense, despite the formidable nomenclature and diagrams. It's really not that hard.

Polyester is the Madison Avenue name for polyethylene terephthalate. The abbreviation is PET. Be careful, though, because PET is not a polyethylene-type chemical; it's a xylene derivative. If you remember that esters usually end in -ate, and are based on the acid from which they are derived, then you can see that the "ester" in polyester is ethylene terephthalate.

Early production of PET involved the condensation polymerization of terephthalic acid (PTA) and excess ethylene glycol in the presence of a catalyst. This process produced a low molecular-weight polyester prepolymer. The final step is to lengthen the prepolymer by heating it to about 525°F under a high vacuum, producing the final polyester. A simplified version of this reaction is shown in Figure 22–8.

Like nylon, PET can be extruded, cooled, cut into chips, and stored for later melt spinning. It also can go directly to the spinneret. The downstream operations of the spinneret are much the same as with nylon.

Unlike nylon which is highly crystalline, PET fibers are amorphous after spinning. They are like the molecules shown at the top of

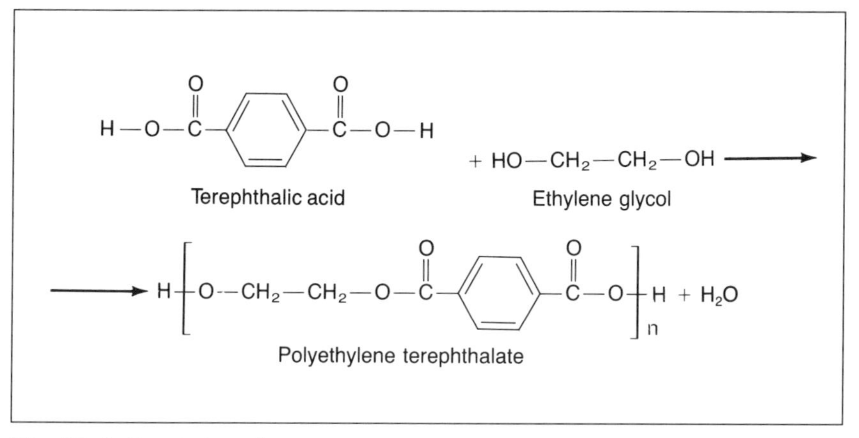

Fig. 22–8 Route to polyester

Figure 20–6. In order to make a usable textile yarn or staple fiber out of PET, it must be drawn under conditions that result in orientation and crystallinity. This is accomplished by drawing at temperatures of about 175 °F with stretch of 300–400%. As with nylon, the conditions of draw (especially the amount) determine the tensile strength and strength and shrinkage properties. Industrial PET fibers, such as those suitable for tires, are more highly drawn.

Other important properties of PET fibers are heat setting (permanent press) and the ability to blend with cotton and wool. By appropriate texturing, various finishes are possible: fur-like deep pile for coats, jackets, bath mats, and soft toys and strong, coarse fiber for tire cords, V-belts, fire hoses, and carpeting. Polyesters are very resistant to degradation by sunlight. Dacron is still the most popular sail cloth because of its weight, its quick drying characteristic, its resistance to stretch and to mold, and its color fastness.

Acrylic Fibers. Acrylic fibers are polymers of acrylonitrile and another chemical. When acrylonitrile is 85% or more of the polymer, the fiber is called acrylic. If there's more copolymer so the percentage of acrylonitrile decreases to 35–85%, the fiber is called modacrylic. Some of the popular monomers used as copolymers are methyl acrylate and methacrylate, acrylamide, vinyl acetate, vinylidene chloride, and vinyl chloride. Dynel is 40% acrylo and 60% vinyl chloride.

These polymers are generally made by the solution polymerization method. The cooking time in the reactor takes 30–60 minutes before the chains are long enough. The basic polyacrylonitrile chain is like the polypropylene chain—it grows along the ethylene backbone. The cyanide grouping is hung over the side.

Melt spinning is not used for polyacrylics because they are sensitive to high temperatures. They actually begin to decompose before they reach melting temperature. Solution spinning is used instead. The dried polymer is dissolved in a polar solvent like acetone or dimethyl formamide (DMF). The spinning mechanics are otherwise the same, except the solvent is recovered as it vaporizes, immediately after the extrusion through the spinneret. Most acrylics are sold and used in the form of staple fiber.

Acrylic fibers generally have good "hand," as it's called in the business (they're soft). They resist creasing, and they are quick to dry. Acrylics have replaced wool in many applications, such as blankets and sweaters. Because of their unique bulking characteristics, they take on the appearance of wool yarn.

Polypropylene Fibers. A small part of the total fibers market (and therefore of this section on fibers) is fiber grade polypropylene. The chemistry for polypropylene fibers is the same as for thermoplastics. The spinning mechanics are the same as that for nylon. Polypropylene fibers are particularly resistant to abrasion and chemicals, and they are lightweight. However, they don't take colors very well, and the materials have low softening points and low resilience (they wrinkle). The major applications for polypropylene fibers are carpet-face fiber and backing (because it's tough) and rope (because it is strong and floats in water).

Chapter XXII in a nutshell...

Thermosets are generally a set of virtually identical, high molecular weight compounds that are cross-linked together by chemical bonds. The cross-linking is what prevents the remolding, melting, or dissolving. Phenolics, epoxies, and polyurethanes are examples.

Phenolic resins are based on phenol and some other comonomer like formaldehyde. Phenolics are widely used in areas of tough duty, and where they don't have to look pretty, such as switchgear, handles, and plywood glue.

Epoxies are made from Bisphenol A and epichlorohydrin. They are tough, resistant to heat, and have good adhesive properties. They can be mixed on-site by adding a catalyst and filler to the already linked Bisphenol/epichlorohydrin which causes the cross-linking and the "set" in thermoset.

Polyurethanes can be made from toluene diisocyanate and propylene glycol, or more complicated forms of these compounds. Polyurethanes are easy to foam either in flexible or rigid form.

Engineering resins are polymers designed to excel in certain physical characteristics. For the most part they are thermoplastics, but they can be thermosets as well. Examples are Nylon 6, Nylon 66, and polycarbonates. Phenolics can be considered engineering resins.

Nylon 6 is made by polymerizing caprolcatam, a derivative of cyclohexane. Nylon 66 is made by the polymerization of adipic acid and hexamethalinediamine. Polycarbonates are made from Bisphenol A and phosgene.

Fibers, like Nylon 6, Nylon 66, acrylics, and polyesters, have to be made from thermoplastics. To create the filaments, the molecules have to be oriented (literally all be lined up in the same direction), either through reheating or by dissolving them.

Exercises

1. What are the ingredients of polyurethane? Is polyurethane a thermoplastic or a thermoset?
2. What's the difference between rigid foams and flexible foams?
3. Which of these are thermosets?

    nylon
    polycarbonates
    bakelite
    epoxides

COMMENTARY FOUR

"I hate quotations.
Tell me what
you know."

■

**Journal, Ralph Waldo Emerson
(1803–1882)**

REVIEW

How do you summarize a summary of a whole body of chemistry like the polymers? You only pick out some tidbits and you can summarize what comes from what with the following table.

The differences in the properties of polymers goes back to the chemical configurations. In simple terms, thermoplastics can be molded because they are long chain molecules that slip if pushed or pulled, especially at higher temperatures. Thermosets are cross-linked, so the long chains stay put under stress, strain, or heat. The fibers get their flexibility and strength when the polymer molecules align during filament formation.

If you've gotten this far (even if you've skipped a couple of pages or chapters), you need to be reminded of the good news. There's an excellent index coming up right after the Appendix. You can use it to refresh your memory on those one or two points that you might forget sometime in the future.

PLASTICS AND RESINS COMPOSITIONS

Thermoplastics

Abbreviation/Name	Monomer
LDPE - low density polyethylene	ethylene
HDPE - high density polyethylene	ethylene
LLDPE - linear low density polyethylene	ethylene and butene-1 or other low molecular weight alpha olefins
PP - polypropylene	propylene
PVC - polyvinyl chloride	vinyl chloride
PS - polystyrene	styrene
Thermosets	
Phenolics	phenol and formaldehyde
Polyurethane	diisocynate and propylene glycol or a polyglycol
Epoxy resin	Bisphenol A and epichlorohydrin
Urea-formaldehyde	Urea and formaldehyde
Polycarbonate	Bisphenol A and Phosgene
Fibers	
Nylon 6	Caprolactam (from cyclohexane)
Nylon 66	Adipic acid and hexamethylenediamine
Polyacrylics	Acrilonitrile
PET - polyester	Terephthalic acid and ethylene glycol

APPENDIX

"Work for the work's sake,
then, and it may be
That these things shall be
added unto thee."

■

Kenyon Cox, 1856–1919

Conversion and Yield

Conversion and yield are often used to describe the efficiency of a plant or a process. Sometimes the semantics get mixed up and one is substituted (erroneously) for the other.

Conversion is defined as the percent of the feed that disappears in a chemical reaction. Take the pounds of benzene, for example, coming out of the ethylbenzene reactor, divide by the pounds of benzene going into the reactor, subtract that from 1.0 and multiply by 100, and you get percent benzene conversion. Conversion is usually measured around a reactor, not around the whole plant. In the case of the ethyl-benzene plant, for example, since there's a benzene recycle, there is virtually no benzene leaving the EB plant. So plant conversion is often not a very helpful concept, but the conversion across the reactor is.

Yield is a more difficult concept, first because it's used in several different ways, and second because one of the uses requires some basic chemistry. In the simplest case, yield refers to the pounds of product leaving a reactor divided by the pounds of feed. In an olefins plant, for example, the ethylene yield is equal to the pounds of ethylene divided by the pounds of, say, gas oil feed. (This definition of yield is more commonly used in refining than in petrochemicals, but then, ethylene plants are usually in refineries.)

The petrochemical and more technical definition of yield refers to the amount of converted feed that actually ends up as product. This is not an easy subject for the "non-technical" person, but here goes.

Definitions

Atomic weights—the relative weights of atoms; hydrogen is the lightest and is defined to be one; carbon is 12 and oxygen is 16, meaning they are 12 and 16 times as heavy as hydrogen.

Molecular weight—the sum of the (relative) weights that make up a molecule; the molecular weight of water, H_2O is 2 + 16 = 18; of methane, CH_4, is 12 + 4 = 16.

Mole—The molecular weight of a compound expressed in grams is the gram molecular weight, usually contracted to the term mole. If the weight is expressed in kilograms or pounds, then it's kilogram moles or pound moles. A mole of water would be 18 grams of water; a pound mole of water would be 18 pounds of water.

The theoretical yield of a reaction can be calculated based on molecular weights. For example, the reaction of hydrogen with carbon dioxide to give methanol:

$$3H_2 \quad + \quad CO_2 \longrightarrow CH_3OH \quad + \quad H_2O$$

$$3(1+1) \quad + \quad (12+16+16) \qquad (12+1+1+1+16+1) \quad + \quad (1+1+16)$$

molecular weights:

$$6 \quad + \quad 4 \qquad\qquad 32 \quad + \quad 18$$

Theoretically, three moles of hydrogen (6 grams) will react with one mole of carbon dioxide (44 grams) to give one mole of methanol (32 grams) and one mole of water (18 grams). In actual practice, a by-product, dimethyl ether, also gets formed:

$$2CH_3OH \longrightarrow CH_3OCH_3 + H_2O$$

Suppose that in an experiment, 22 grams of CO_2 reacted with excess H_2 to give 12 grams of methanol and some dimethyl ether. To determine the ACTUAL YIELD of methanol based on the CO_2 reacted:

Step 1 — Determine the moles of CO_2 reacted and the moles of CH_3OH produced:
22 grams of CO_2 is half of a mole of CO_2 (22/44) and 12 grams of CH_3OH is 0.375 moles (12/32).

Step 2 — Theoretically, one mole of CO_2 will give one mole of CH_3OH; so a half a mole of CO_2 gives half a mole of CH_3OH.

Step 3 — Divide the actual moles of product by the theoretical moles of product produced.
0.375 / 0.500 = 0.75
The actual yield is 75%.

Note that the yield is not the weight of the product (12 grams) divided by the weight of the feed (22 grams). Using molecular weights as an adjustment helps track what happens chemically to the molecules so that the chemist or chemical engineer can determine whether to work on improving the process.

EXERCISE ANSWERS

"Putting on the spectacles of
science, in expectation of finding
the answer to everything looked at,
signifies inner blindness."

■

The Voice of the Coyote
J. Frank Dobie, 1888–1964

Chapter I

1. This is a mixed bag of correspondence:

   Paraffins have the formula C_nH_{2n+2}.

   Olefins have the formula C_nH_{2n}.

   Aromatics, or at least benzene, have the formula C_nH_{2n-6}

   Saturates are paraffins.

   Examples of unsaturates are the butylenes.

   Examples of isomers are the three xylenes.

   Examples of cyclics are benzene, xylene,

   and toluene.

2. There are 3 isomers of pentane:

$CH_3CH_2CH_2CH_2CH_3$

normal pentane

$CH_3CH_2CHCH_3$
|
CH_3

iso-pentane

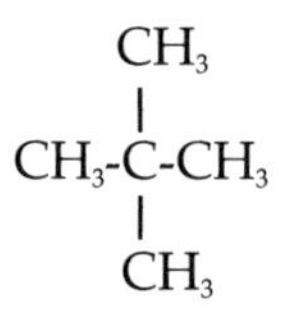

neo-pentane

3. $-C_2H_5$. Only five hydrogens, not six. One hydrogen has to be dropped from the configuration to reflect the bond that connects the group to some other group.

4.

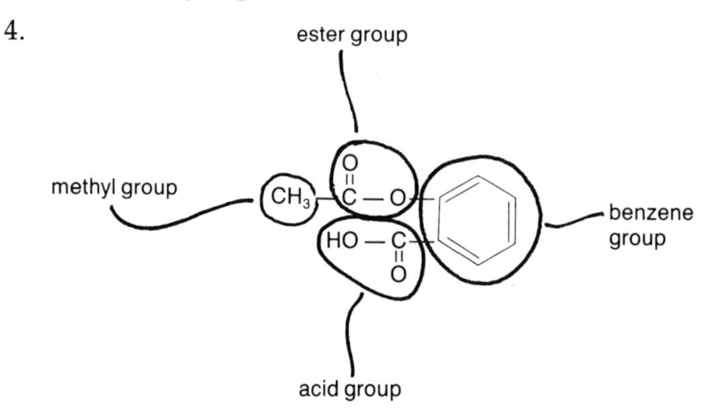

Chapter II

1. naphtha - cat reformer - reformate
   coal - destructive distillation - benzene
   toluene - hydrodealkylation - benzene
   gas oil - olefin plant - benzene
   reformate - solvent extraction unit - benzene

2. Figure out the value of the pounds of toluene and hydrogen in, plus the operation cost, and compare it to the value of the benzene out:
   — 500,000 times 7.21 times 0.20 = $721,000 for toluene

— from Table 2–2, you need 27 lbs. of hydrogen for every 1200 lbs. of toluene, or 81,112 lbs.
— 81,112 times 0.40 is $32,445
— total raw material cost is $753,445
— benzene yield is 1000 lbs. for every 1200 lbs. of toluene, or 3,048,780 lbs.
— 3,048,780 times 0.24 = $731,703, which is less than the raw materials cost.

Answer is: sit tight and wait for the price of benzene to go up.

3. Platinum catalyst
4. Reformate, the product of reforming, has aromatics in it. Raffinate is the result of running reformate through an extraction unit, taking the aromatics out.
5. The feed is regular coffee, the solvent is CH_2Cl_2, the extract is caffeine, and the raffinate is de-caf coffee. (Enjoy your next cup of raffinate.)

Chapter III

1. Benzene, toluene, ortho-xylene, meta-xylene, para-xylene, and ethylbenzene.
2. toluene - hydroalkylation - benzene
   mixed xylenes - adsorption - para-xylene
   toluene - disproportionation - mixed xylenes
   mixed xylenes - cryogenic distillation - para-xylene
3. Sure, because the freezing points of benzene, toluene, and meta-xylene are 167.2, 231.4, and 282.4°F, far enough apart that distilling by freezing will work. The more usual method is fractional distillation, which is cheaper in both capital cost and operating cost.
4. e.

Chapter IV

1. When hydrogen atoms are added to the benzene molecule (hydrogenation), the chemical reaction gives off heat (exothermic).
2. a. six hydrogen
   b. catalysts
   c. nylon

3. You could pay almost 28.5 cents/lb.

   1000 @ 30 cents + 9 @ 0 cents =

   944 @ X cents + 65 @ 40 cents + 1009 @ 0.5 cents

   X = 28.5 cents

   (the 1009 is the total feed, 944 + 65)

4. The difference between eggs and chickens and cyclohexane and benzene is that, "for sure," Mother Nature made both cyclohexane and benzene at the same time.

Chapter V

1. If you crack 1.25 billion pounds of propane or 2.778 billion pounds of gas oil, you'll get at least 500 million pounds of ethylene and more than the minimum amount of propylene:

   500 divided by 0.40 ethylene yield from propane
   = 1250 million pounds of propane feed

   200 divided by 0.18 propylene yield from propane
   = 1111 million pounds of propane feed

   500 divided by 0.18 ethylene yield from gas oil
   = 2777 million pounds of gas oil feed

   200 divided by 0.14 propylene yield from gas oil
   = 1428 million pounds of gas oil feed

   Cracking the propane gives 1.25 billion times .01 or 12.5 million pounds of butadiene; the gas oil gives 2.778 times 0.04, or 111.12 million pounds of butadiene.

2. Because it's inside a tube and does not come in physical contact with the fire in the furnace.

3. Hydrogen, methane, ethane, ethylene, acetylene, propane, propylene, butane, iso-butane, butylenes (butene-1, butene-2, iso-butylene), butadiene.

4. Since there's plenty of propylene around refineries, olefin plants didn't really have to be built to make propylene, only ethylene. But a petrochemical that has an annual demand of 13 billion pounds can hardly be considered second class.

Chapter VI

1. If they're in the chemical business, and the price of butylenes is suffi-

ciently lower than butadiene, which it often is, it makes sense to build a dehydrogenation plant to make butadiene from butylenes. But if they are a petroleum refiner and have no petrochemicals business, then they want to get the butadiene out of the C_4 stream before it goes to the alkylation plant. Butadiene will cause the process to foam up, lowering conversion and making a general mess.

2. c. and d., depending on the price of butane and butylenes. In a. and b., butadiene is a by-product, and the availability can not be varied, for all practical purposes. And Sal went into Chapter 11 last year.
3. Extractive distillation is used to remove butadiene from a C_4 stream; fractionation can be used to separate out butene-1; adsorption is also sometimes used to separate out butene-1; polymerization is sometimes used to pull out the iso-butylene; dehydrogenation can be used to convert some of the butylenes and normal butane to butadiene; and alkylation is used to convert the butylenes to alkylate.

Chapter VII

Net cost of the cumene (benzene + propylene + operating cost):

681 times 1.60/gal divided by 7.19 lbs./gal = 151.54
367 times 0.25/lb. = 91.75
1048 times 0.20/lb. = 209.60
452.89

452.89 divided by 1000 equals $ 0.453 per lb. of cumene.

Net cost of the phenol (cumene + oxygen + plant operating cost - acetone - by-product values):

1000 times 0.453/lb. = 453.0
300 times 0.500/lb. = 150.0
725 times 0.25 = 181.25
-442 times 0.60 = -265.2
- 99 times 0.10 = - 9.9
509.15

Divide by 725 to get the breakeven cost of phenol. Add $0.02 to get the sales price of $0.722/lb.

2. Cutthroat forgot, once again, that when you produce all that by-product acetone (30 million lbs.) on the market, its market price is not

likely to stay at $ 0.60/lb., and that changes the calculation of the breakeven cost of the phenol.

Chapter VIII

1. ...cumene...styrene
2. Acetone could be considered as by- or co-product of cumene manufacture. Propylene oxide is a by- or co-product in one of the styrene processes.
3. Benzene 1000.0 lbs.
   to ethylbenzene 1345.89 lbs.
   to styrene 1188.95 lbs.
   using 395.35 lbs. of ethylene.

Chapter IX

1. Quench is hot VC/HCL plus cool EDC. Oxychlorination is HCl, O_2, and C_2H_4 to EDC. Pyrolysis is EDC to VC and HCl. Chlorination is C_2H_4 and Cl_2 to EDC.
2. $2CH_2=CH_2 + Cl_2 + 1/2\, O_2 \rightarrow 2CH_2=CHCl + 2H_2O$
3. Chlorination is the addition of chlorine to a compound; oxidation is the addition of oxygen to a compound; oxychlorination is the addition of them both. Did you ever think you'd be comfortable with a word like oxychlorination?

Chapter X

1. epoxide — EO
   silver bullet — silver oxide
   epoxide ring — a molecule under stress
   antifreeze ingredient — EG
   EO by-product — CO_2
   EG by-product — heavy glycols
2. The discovery and commercialization of the direct oxidation route using silver oxide as the catalyst.
3. Ten million gallons of EG:

= 10 mmg EG
= (10 x 9.3 lbs./gal.) mmlbs. of EG
= (10 x 9.3 x .816) mmlbs. of EO
= (10 x 9.3 x .816 x .850) mmlbs. of ethylene
= 64.5048 mmlbs. of ethylene required, theoretically

Chapter XI

1. epoxidation
   chlorohydrin...indirect oxidation...direct oxidation
   TBA...styrene...dipropylene glycol
2. Bad economics: chlorine disposal, energy intense
3. See figure 8–6.

Chapter XII

1. The oxidation and the partial oxidation method, and the CO_2 from an ammonia plant in a reaction with methane and water all produce different ratios of CO and H_2. In addition, CO_2 can be removed by solvent extraction. So the trick is to use two or three of these processes to get the $CO:H_2$ ratio to about 1:2.
2. Ammonia plants produce an otherwise worthless by-product, CO_2, which can be uses as a co-feed to a methanol process.
3. Methanol is used primarily to make MTBE, acetic acid, and formaldehyde, and is used as a gasoline blending component or neat as an automotive fuel.

Chapter XIII

1. Because ethyl alcohol forms an azeotrope with water that is a constant boiling mixture, i.e., both the ethyl alcohol and the water , in a ratio of 95/5, boil together at a temperature different than either separately. Other examples in this chapter are the ternary azeotrope, ethyl alcohol - water - benzene; iso-propyl alcohol - water; and DIPE - isopropyl alcohol - water. An azeotrope mentioned earlier is MEK - water - toluene raffinate used for toluene extraction.
2. Because "like dissolves like." The -OH group of IPA is a major part of the molecule, as it is in water, H-OH. That makes IPA "like" water. But

in normal hexyl alcohol, CH_3-CH_2-CH_2-CH_2-CH_2-CH_2-OH, the -OH group is not a dominant part of the molecule, and "unlike" water.

4. Calculation as follows:

   Fermentation cost –

   $2.50/bushel divided by 2.6 gallons/bushel
   plus $0.50/bushel
   equals $1.46/gallon

   Petrochemical process –

   1 gallon of ethyl alcohol = 6.58 lbs.
   which requires X lbs of ethylene to produce.
   1000/640 = 1.56 lbs. of ethyl alcohol/lb.
   ethylene
   X = 6.58/1.56 = 4.22 lbs. ethylene needed to
   produce 1 gallon of ethyl alcohol

   so, fermentation cost $1.46/gal.
   less p.c.plant op. cost .30
   gives 1.16
   divided by 4.22 lbs. ethylene/gal ethyl alcohol
   gives $0.275/lb. ethylene

Chapter XIV

1. a. formalin
   b. formaldehyde
   c. formaldehyde ... formaldehyde
   d. acetaldehyde or formaldehyde
   e. CO_2 or water
   f. aldehyde
   g. acetaldehyde
2. The dictionary (at least one late edition) says that an oenophile (oeno-, wine; philos, love) is a wine aficionado. And wine that has some aldehydes in it that never converted to ethyl alcohol is bad wine.
3. ethane ⟶ ethylene ⟶ ethyl alcohol
   ↓
   acetaldehyde
   ↓
   acetic acid
   ↓
   $CO_2 + H_2O$

Chapter XV

1. The only structural difference between the aldehydes and the ketones is in the position of the signature group, the double bonded oxygen. For ketones, it is always attached to a carbon in the middle of a hydrocarbon chain. For an aldehyde, it is always attached to a carbon at the end of the chain.

2. Using the material balances for an IPA plant, an acetone plant and an MIBK plant,

   1000 lbs. of MIBK needs 1160 lbs. of acetone and 20 lbs. of hydrogen;

   1160 lbs. of acetone needs 1160 times 1.158 = 1343 lbs. of isopropyl alcohol;

   1343 lbs. of isopropyl alcohol needs 1343 times 0.9 = 1209 lbs. of propylene.

   So breakeven cost of the propylene is:

1000 times $0.45	450
less hydrogen cost	
20 times $0.20	4
less plant tolling costs	
1000 times $0.05	50
1160 times $0.05	58
1343 times $0.05	67
equals	271

   divide by 1209 lbs. of propylene

   breakeven cost of propylene = $0.224/lb.

3. a. formaldehyde
   b. acetone
   c. formaldehyde
   d. MEK
   e. nickle
   f. zinc oxide
   g. isopropyl alcohol
   h. Wacker process
   i. methyl methacrylate
   j. butylene
   k. MEK
   l. methyl isobutyl carbinol
   m. palladium chloride

Chapter XVI

1. a. In an acid, the signature group has a hydroxyl group, -OH attached to the carboxyl carbon. In an ester, it is an -OR attached to the carbon where R can be an alkyl group like $-CH_3$, $-C_2H_5$, etc., but never a hydrogen atom.
   b. Where the acid signature group has a hydroxyl attached to it, the aldehyde has a hydrogen attached instead.
   c. An acid anhydride is two acid groups joined together by removing a water molecule, resulting in that wrap-around signature group.
   d. Oxidize an acid some more and you end up with carbon dioxide (and water).
2. An acid has a double-bonded oxygen where an alcohol has two hydrogens.
3. Both of them can easily be converted to an anhydride.
4. Nylon 66 ....phenol....vinyl chloride....phthalic anhydride....styrene.
5.

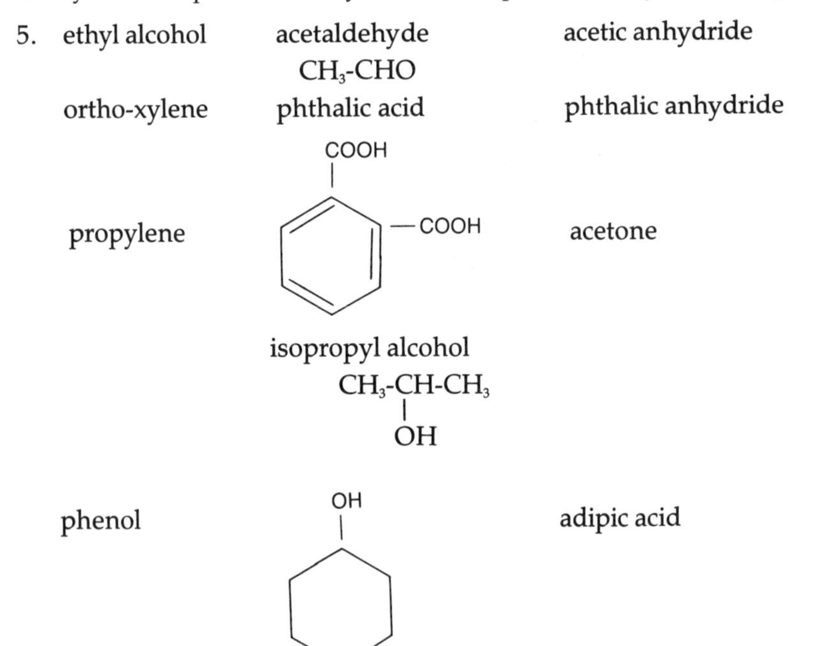

Chapter XVII

1. Acrylonitrile and acrylic acid can both be made out of propylene. Acrylates and acrylic acid can both be made out of acrylonitrile.
2. In a fixed bed process, the feedstocks move over, around or through the catalyst, which stays in one place all the time. In a fluidized bed process, the catalyst used behaves like a fluid as moves through the reactor along with the feedstock. It is separated at the end of the reaction. Because the catalyst is so small and there is plenty of it, and because it gets surrounded by the feedstock, it effectively has a very large surface area per pond of catalyst. That facilitated the reaction, what with the feedstock coming into such intimate contact with all that catalyst. Th major disadvantage of the fluidized bed reactor is the loss of the catalyst into the product stream during the catalyst recovery phase.
3. Vinyl chloride, vinyl alcohol, styrene, acrylic acids, acrylonitrile, the acrylates, propylene, acrolein.
4. Esterfication starts with two compounds, like an alcohol and an acid, and ends up with two compounds, an ester and water. The water is formed by each of the starting compounds giving up one or two atoms to form the H_20. Dehyrdration starts with one compound like alcohol and ends up with two, generally an olefin and water. The water is made up of atoms given up by the original compound.
5. One of the olefins, propylene, is a commercial staring point for all those compounds. Commercial routes from ethylene, butylene, etc., have not been developed.

Chapter XVIII

1. Acetic anhydride (acetic acid), phthalic anhydride (ortho-xylene), maleic anhydride (butane or benzene). Phthalic anhydride has a benzene ring as its unique feature, with its three double bonds; maleic anhydride has its single, but reactive, double bond. Acetic anhydride

has only the anhydride signature group. It has neither a closed ring structure nor a carbon-carbon double bond.

The most reactive of the three is maleic because of the combination of the anhydride group and the highly reactive double bond.

2. Cobalt acetate is used to make acetic acid out of butane; palladium chloride/cupric chloride are used to make MEK out of butylene; and lo and behold, to make phthalic anhydride out of ortho-xylene or naphthalene, you use vanadium pentoxide.
3. The two most important factors are the cost of the feedstock and the yield of MA from each.

Chapter XIX

1. In the higher alcohol process, the displacement is effected by oxidizing the trialkyl aluminum and then hydrolyzing, to form aluminum hydroxide and the linear alcohol. For alpha olefins, ethylene is used to displace the alpha olefin.
2. Because the addition is done by adding ethylene, and the carbon count in that comes in two's.
3. Back in Chapter III Murphy's Law stated that anything that can happen will happen. The control of the rate of ethylene addition is not all that good. So the rates vary from molecule to molecule, and out comes a distribution of alpha olefins.

Chapter XX

1. Thermoplastics can be remolded several times by applying heat and/or pressure. Once thermosets are cured they can not be changed in shape by heat or pressure.
2. Right, there's only three: initiation, propagation, and termination.
3. Bulk, suspension, solution, and emulsion polymerization.
4. Thermoplastics—because they need to be re-oriented by pressure or temperature when they are drawn into fiber.
5. a. Homopolymer of formaldehyde (CH_2O)
   b. Copolymer of adipic acid and ethylene glycol
   c. Homopolymer of vinyl chloride

Chapter XXI

1. When a polymerization uses two or more different types of monomers, it is called copolymerization. Examples are styrene acrylonitrile, LLDPE (ethylene and butene-1), and polyester (ethylene glycol and maleic anhydride).
2. The atactic polypropylene is soft and rubbery and doesn't have many useful characteristics.
3. Start with ethylene and chlorine and make ethylene dichloride; then crack it and make vinyl chloride; then polymerize it and make polyvinyl chloride.

Chapter XXII

1. Typically toluene diisocynate and propylene glycol. Polyurethane can be either a thermoplastic or a thermoset depending on the monomer used. If propylene polyol is used in place of propylene glycol, there are more sites for cross-linking and a thermoset will result.
2. Rigid foams generally have closed cells; flexible foams have open cells so the air can escape during flexing.
3. Nylon isn't.

INDEX

"There is a time for many words, and
there is also a time for sleep."

Homer c. 700 b.c.